AN ILLUSTRATED HISTORY OF CRANES

FGL
FGL

ACKNOWLEDGEMENTS

I would like to thank all those who have so kindly helped me with this book in whatever way; all those who sent a photograph or photographs, illustrations or drawings and histories of companies.

I especially want to thank the following: Mr Neil Richardson and his kind colleagues at British Ropes Ltd, Carr Hill, Doncaster; Mr Lynn Doylerush at the Boat Museum; Gloucester Waterways Museum; Prof F. W. Hampson, Clarke Chapman Marine; DEL Crane Services (Kent); Mr Martin Ainscough, Ainscough Crane Hire; Mr Peter Gilewicz of Marion Power Shovel Co (USA); Mr Les White of Eastleigh; Mr Colin Chambers of Cheltenham; Mr Tony Swaine; Mr Gordon R. Fairbairn from Ipswich; Mr Roger M. Elliott of BSP International Foundations Ltd; Mr M. Finch, MD, Casagrande, UK; Mary A. Sellers and Norman Hargreaves of PPM Cranes Inc (USA); Mr Dieter Jebens; Mr P. C. Milward; Samantha Smith, PR Assistant, ABP Southampton; Mrs Pauline Hall; Colin Baker of *Cranes Today*; Brian Chantler of Henry Cooch & Son Ltd; Peter Simms and Peter Christie (Bridge Trust), Bideford; Concept Staffing (the two Lisas); Mr Philip Peacock; Mr Dave Barnaby; Colin Self; Capt R. T. Arnold (Retd) and the staff of the Royal Engineers Library, Chatham; Sara Sharpe and John Lyle Carter (Plymouth Development Corporation, Royal William Yard); and to everyone else who assisted, not least my wife Caroline for her unceasing patience.

FOREWORD

In my previous book for Ian Allan Publishing — *An Illustrated History of Excavators* — it was relatively easy to make good use of chapters. For example, cable shovels, draglines, hydraulic excavators and bucket wheel type excavators were separated from other forms of excavator.

In this book, however, it has not been so easy. A chapter, say, on steam cranes can take several forms — are they mounted on crawler tracks, and used predominantly in the construction industry, are they mounted on railway lines, used in the railway or in the cargo handling industries at docks and ports? Is a crawler crane when mounted on a pontoon or a barge classified as a marine crane or is that the category exclusive to cranes permanently mounted on a ship?

So cranes do tend to cross a number of boundaries. The heading, therefore, may include construction, railway, marine and dockside cranes all under the one roof. I have also attempted to enter the photographs in a manner beginning with the earliest through to the most recent, or the smallest to the largest. This, however, has been possible in the majority but not every case.

To those who like looking at pictures of machinery, as in the case of the hundreds who have bought copies of the *Excavators* book, I feel there is much to be gained from adding this book to a collection, including: information, photographs (many of which were not previously available in book form) and a guide to the cross section of users and where possible the manufacturers of cranes and the historical facts. While every endeavour has been exercised to be as accurate as possible over dates, machine identity and other basic facts, errors do occasionally occur.

To those companies who would wish their early or more recent products to have been included, so do I. You may well have been contacted but overlooked my request for photographs and historical facts. If you were not contacted, we are both the loser, as is the reader, for which I can only apologise.

Finally, can I pay tribute to all those who have contributed photographs, whether singly or in large numbers. You have all been the reason this book has appeared for the enjoyment of everyone. My sincere thanks to you all.

Hinton J. Sheryn

INTRODUCTION

When the Arndale Shopping Centre in Manchester suffered the horrific IRA bombing in 1996, a high capacity top-of-the-range truck-mounted crane was brought in to lower the damaged bridge and to help make surrounding buildings safe. When trains collide, buses overturn, or aircraft crash, the crane is often on its way to the scene in the wake of the emergency services. Cranes have been around for a very long time. Their future is assured while construction, repairs, demolition, accidents and progress persist.

In this book, I have attempted to feature a cross section of a variety of cranes. However, to feature all makes and all types would take years and years of difficult research. The book would finally be the size of a complete volume of the *Encyclopaedia Britannica*. Some of the cranes I have not managed to feature include a skip lorry, a concrete pump, and lorry-mounted hydraulic cranes as used by builders' merchants, public utility companies, engineering suppliers and timber hauliers. Logging tractors and cranes, aerial ropeways, and the small cranes fitted inside the rear of some cars to lift a wheelchair into the car for its disabled user, are other types which will have to await a future book.

There are specialist cranes to open huge furnace doors at steelworks and foundries, others to lift the great ladles full of molten iron or copper, or even gold. There are cranes to lift heavy parts, such as electric motors aboard huge excavators, such as the immense draglines and shovels at work for the world's surface mining industry, cranes to lift massive tyres on and off the new generation of 300-ton dump trucks and in other cases to handle tyres on scrapers or large-wheel loaders and bulldozers. There are Straddle Carriers employed on most modern docks to handle containerised cargo. There are hundreds of access platforms, from which personnel can work at otherwise difficult heights. There are special cranes used at vehicle repair garages to lift components such as gearboxes or engines. There are the breakdown trucks which have, themselves, a unique crane.

Talking of engines, specially adapted cranes are used to repair and remove or install the massive engines and turbines on board ships, including aircraft carriers, ocean-going liners and bulk tankers. Another crane, which does not readily spring to mind, is the helicopter, which is used to hoist equipment and personnel in both military and civilian scenarios. The North Sea oil industry has regular helicopter missions to its advantage, as has the odd church requiring a weathervane or flagpole to be installed using aerial lifting methods.

Robots in car factories and on assembly lines are, themselves, a form of crane. Similarly the use of robots in laboratory conditions, where total cleanliness prohibits personnel from working with exposed skin for fear of infection of either the personnel or the objects being handled. The installation and removal of fuel rods at nuclear power plants around the world or on nuclear submarines is another requirement for specialist cranes.

There are, and have been, countless other examples, so forgive me if few of these specialist fields have been covered in this book. A simple pulley, perhaps working from a scaffold to hoist a bucket, or a power winch for similar work, or a chain block in an engineer's workshop or a garage, could have been included. Most of the readers of this book will have seen many such gadgets in daily use, but early steam cranes, 1,000-ton capacity crawler cranes or 6,000-ton capacity marine cranes or tower cranes stretching almost a mile into the sky cannot easily be imagined.

A HISTORICAL BACKGROUND

Cranes have been around for more than 3,000 years. Early users included the ancient Chinese, who used crude lifting devices to load and unload their cargoes of spices, silks and other products from the classic boats traditionally known as junks. Using ropes made from papyrus leaves, these were able to assist the heavy work involved with manhandling large quantities of material.

Through the centuries from these simple beginnings, the crane has become as important a part of man's development as the wheel or indeed the boat itself. Nevertheless many things from the past still remain mysteries, such as the methods used to construct the Pyramids, Stonehenge, or the Great Wall of China. Using huge stones weighing, in many cases, several hundred tons, it is hard to imagine quite how these could be hoisted to their final resting places without the help of great cranes similar to those in use over the past 150 years or so.

In modern times, massive jetties, ports, docks, breakwaters, dams, cities, cathedrals, churches, bridges and monuments have all been built with the aid of some form of mechanical lifting device. They may have been huge tripod-type shear legs, with hand (or later steam) operated pulleys or winches. They may have been driven by animals, or even men in cages as has been the case, certainly in Germany (and doubtless elsewhere), where prisoners used to walk inside large wheel-cages (similar to those we see in hamster cages) serving like capstans, thereby hoisting or lowering a heavy load from the long wooden boom.

For many centuries it was the wooden boom (or jib) which predominated, until the increased use of iron and steel in the 18th and 19th centuries.

Many hoists and cranes were reliant on hand power through reduction gearing to power the variety of crane types then available. Soon, however, the greatest advance in crane technology was to emerge — steam.

With this new-found power source and the use of heavy gauge iron and steel, the crane was able to grow in both stature and strength. As a consequence, the advancing railways, bridges, docks and ships, themselves using many thousands of tons of steel and iron, were all able to make good use of the powered crane.

With the beginning of the industrial revolution, bigger and better ships were required to transport goods and people to and from all parts of the world navigable by sea. Thousands of cranes were required to build the ships and far more were required around the world to load and unload their cargoes. Anyone who was prepared to invest their capital and engineering skills in the building of cranes was sure to be in for a bonanza.

Some of the early pioneering companies are still world leaders in the crane-making industry. Others have lost their origins through acquisition or mergers, while others have disappeared altogether, although just a few of their early products might well have outlasted the manufacturers themselves. Some of the

Right: **The original 'luffing boom' crane. A vertical wooden pole, supported by ground-anchored guy ropes. Angled pole controlled by: 1) adjusting pole base knuckle swivel; 2) manual luffing ropes reeved through double and triple Fairlead blocks, led back to ground-anchored winch. Chris Miller, one of the country's oldest crane hire/crane engineering companies, erected this lifting device in the early 1900s.** *Chris Miller Ltd*

best remembered names have been lost as world leading makers of cranes over a period from the 1960s: Priestman Brothers of Hull, Thomas Smith of Rodley, near Leeds, Butters Brothers of Glasgow, Marion of Ohio (USA), Ransomes & Rapier of Ipswich, John Allen of Oxford, Coles Cranes of London and Sunderland, Jones Cranes of Letchworth, Hertfordshire, Baldwin-Hamilton-Lima, of Lima, Ohio, USA and Ruston-Bucyrus of Lincoln.

Taking their place are such names as Tadano, Kato, Liebherr, Hitachi etc. Marion is now producing very large mining shovels and draglines, while Ransomes & Rapier was absorbed in Bucyrus (Europe) Ltd and makes a few large walking draglines from time to time, mainly these days in India for their coal mining industry.

Ruston-Bucyrus produces the VC range of hydraulic long range backacter/dragline, once manufactured by Priestman Brothers of Hull. Hydraulic cranes are still designed and built by RBI (Ruston-Bucyrus International) Lincoln Works, which once turned out cranes and excavators destined for construction, mining and quarrying sites or irrigation schemes the world over. It would be no exaggeration to state that Ruston-Bucyrus, and its US parent company, Bucyrus-Erie, were the best known makers of cranes and excavators with machines hard at work in virtually every country in the world. Many are still at work, often after 50 or more years of faithful service.

Above: **Chris Miller Ltd supplied this early Fordson tractor to remove the dressed logs from this woodland area by power winching and towing. One three-leg lifting frame was erected with the hoisting facility attached to the tractor, for mechanical lifting through a horizontal winch operated from the rear of the tractor. The winch/hoist rope was led through a multi-sheave Fairlead block, mounted under the three-leg lifting frame.**
Chris Miller Ltd/Martin Ainscough

One of the earliest known crane makers, certainly in the UK, was that of John Padmore of Bath. He produced cranes in the 1730s for work in the surrounding quarries. It might be no coincidence that Bath has since produced one of the world's best known crane-making companies, Stothert & Pitt. Over the many years since the early to mid-19th century, cranes from Stothert & Pitt have seen service on railways and in ports throughout the world. Some of them have been very small hand-operated hoists, others among the largest cranes ever built, some being as easily recognisable in some towns as is the Blackpool Tower or Nelson's Column.

Right: **Two sets of three-legs, fitted with one set of chain blocks suspended under each of the two lifting arrangements. Supplied by Chris Miller Ltd, who erected, attached slings and hoisted this early mobile mixer, prior to loading it for road transport away from the site. This would have been the predecessor of the 25-ton crane, lifting beam and attachments.**
Chris Miller Ltd/Martin Ainscough

Above: This steam crane was being demonstrated at the International Exhibition in 1862/63. In common with many cranes of this era, before the common use of wire rope, it featured a heavy duty steel chain. Note in the background a pulley block operated from a tripod, known as shear legs, a very useful lifting device and much used in the 19th and early 20th centuries in ship building and heavy construction projects. Smaller examples were geared with hand winches, while larger devices made use of steam powered winches. *Courtesy Royal Engineers Library, Chatham.*

Left: The Ruston Proctor patent Steam Navvy was developed in the 1870s. This example, however, was undergoing testing upon being rigged as a crane. This machine was mounted on railway lines in common with almost all mobile cranes of this period. *Derek Broughton & Colleagues RBI, Lincoln.*

Below left: An early hand-operated crane on a railway station near St Austell, Cornwall; the maker's details are somewhat obscure but it will be similar to hundreds of little cranes made, often by local engineers in the locality, during the 19th century and perhaps occasionally since. *English China Clay International*

Left: **This rail-mounted steam crane from Thos Smith of Rodley near Leeds had progressed from using chains to wire ropes as a means of hoisting the load and derricking the boom in or out.**
Royal Engineers Library, Chatham.

Below: **From 1919, when this link-belt locomotive crane was built, steam cranes were being built at an astonishing rate by literally dozens of manufacturers in the United States, the UK, Germany and elsewhere.**
Author's collection

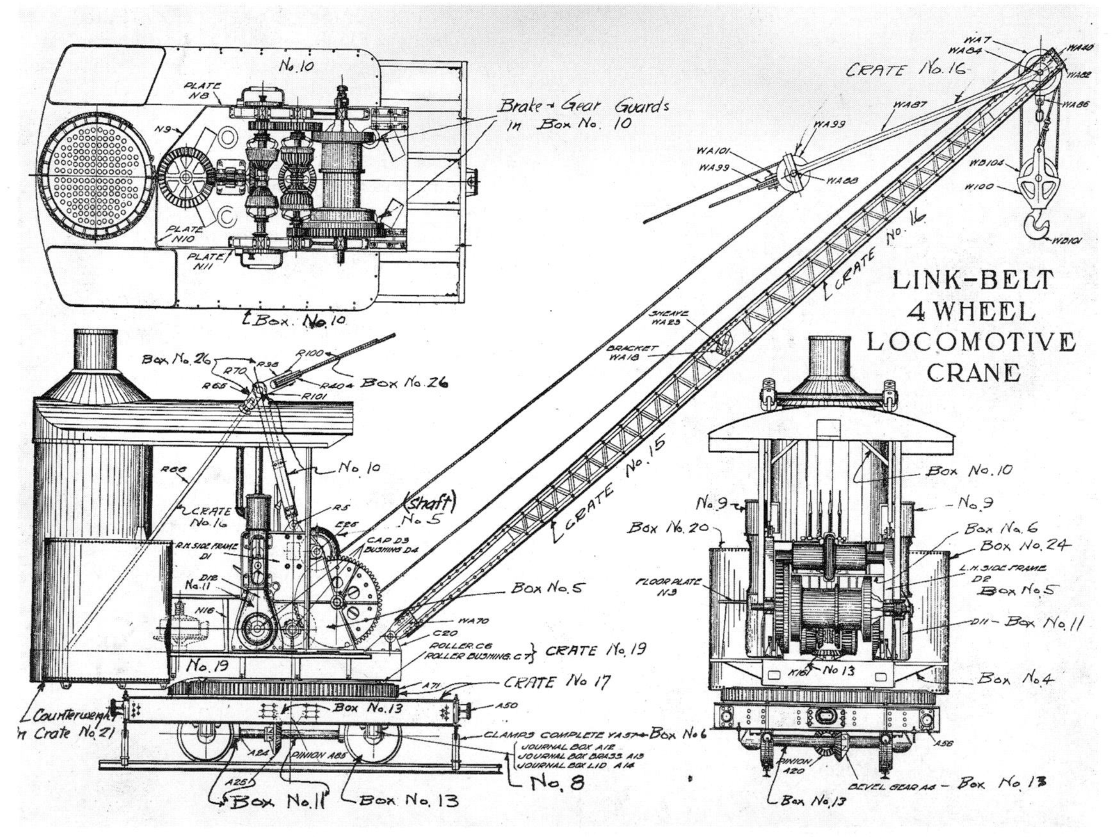

Right: A 12-ton steam crane constructed by John H. Wilson & Co of Birkenhead around 1880. Wilson was best remembered for his Patent Steam Navvy, some of which found themselves hard at work on the Manchester Ship Canal, while other later machines went to work in the ironstone mines and quarries of the Midlands (mainly in the Corby area).
From Bullivant & Co's catalogue, c1902. Author's collection

Left: Grafton Cranes of the Vulcan Works in Bedford was established in 1880. Formerly Mr E. A. Grafton had served his apprenticeship alongside other independent crane makers such as Taylor & Hubbard and Henry J. Coles (of Coles Cranes). All had followed in the footsteps of their mentor, C. J. Appelby, who was thought to be one of the founding fathers of steam railway cranes. Grafton's Standard Steam Cranes included the 2- and 3-ton rail mounted models, a 5-ton model, 7- and 10-ton permanent way steam cranes and fixed steam cranes with or without their own boilers, including 10-ton capacity models. *Courtesy Peter Milward, Maidstone*

Left: Although made by Ransomes & Rapier of Ipswich in Suffolk and supplied to the Manchester Ship Canal Co, this machine was known as Stoney's Patent Tipping Crane in view of the device designed and used to tip the skips, c1889/1890.
From Thos & Wm. Smith's catalogue, c 1900

Below: Also from Ransomes & Rapier was this massive Titan crane built for the Madras Harbour Works in India in the late 1880s. It was also steam powered.
From Thos & Wm. Smith's catalogue, c 1900

Above: **This little waterborne steam dredger, *Perseverance*, is in fact based on a Grafton steam crane. It is working here in Swan Cutting on the Basingstoke Canal in 1976.** *Dieter Jebens*

Above left: A 37-ton capacity steam crane built by Ransomes & Rapier for the Great Eastern Railway in 1892. It was mounted on 12 railway wheels, each provided with springs so as to equalise the pressure on the quay, from where it would operate. It was capable of raising 37 tons at a radius of 36ft.
From Thos & Wm. Smith's catalogue, c 1900

Above: A 16-ton capacity permanent way locomotive crane built by Thos Smith for the North Eastern Railway in the 1890s.
From Thos & Wm. Smith's catalogue, c 1900

Left and below left: Jeremiah Booth was once a partner with Thos Smith of Rodley, until he decided he could build his own range of steam cranes from his factory at the Union Crane Works in Rodley, near Leeds. Their first hand-operated cranes were manufactured in 1840. Eventually the company became Joseph Booth & Brothers, still operating from the same factory, but had been absorbed by Clyde Crane & Booth Ltd. As late as 1949 they were still manufacturing and marketing a range of steam railway cranes. Two of the cranes then available are depicted here. Reproduced from Booth Rodley Type 8/12 Steam Crane catalogue. *Author's collection*

Above: This little steam crane was built by William Balmforth of Rodley near Leeds, who had once, with Jeremiah Booth, been a partner with Thos Smith. Balmforth, like Booth, decided he too could produce his own cranes. This crane was used at the Kirkstall Forge in Leeds until it was sent for scrap in the 1930s. It was rescued by the National Waterways Museum of Gloucester and with plenty of tender loving care has been fully restored by the Dorothea Engineering Co.
National Waterways Museum, Gloucester

Centre left: This very early wooden wharf crane from the Tardebigge Yard on the Worcester and Birmingham Canal rests on a cast iron base. The woodwork was probably replaced in 1944. It was, however, probably made at least 100 years previously.
Gloucester Docks Museum

Left: A rail-mounted, hand-operated crane made by R. C. Gibbins of Birmingham had been used in Sharpness to remove baulk timbers for maintenance of the wooden lock entrance piers.
National Waterways Museum, Gloucester

Right: This little crane is hand operated but somewhat more mobile, being mounted on steel wheels. Its date of origin and the manufacturer are unclear. Such cranes were made throughout the late 19th century and for some years after the turn of the century. *Royal Engineers Library, Chatham*

Below: There were literally thousands of hand-operated cranes produced during the late 18th century and throughout the 19th century. Certainly as far as the UK was concerned, almost every major town had a crane works. Some were pedestal cranes designed to operate from a steel or cast iron plate embedded in the ground, others were mounted on rail-mounted trolleys to give them access along the side of the quay to a line of boats. This crane is one such mobile crane fitted with a moveable counter-weight. It has a maximum lifting capacity of 3 tons. *Boat Museum, Ellesmere*

Bottom: A view of the mills from the docks at Ellesmere Port, where an early steam crane is waiting for its next cargo vessel to arrive. *Boat Museum, Ellesmere*

Above right: Steam cranes aplenty lend a hand on this construction project which is believed to be the construction of the Manchester Ship Canal. Cranes by Thos Smith, Wilson and Priestman were among the famous names employed on this very big excavation. *Boat Museum, Ellesmere*

Right: Three hydraulic wall-mounted cranes were used to lift those huge sacks of grain into the mills and the flour back onto waiting barges in 1908.
Courtesy D. L. MacDougall/Edwin Pratt, Author, British Canals

Far right: Another little wooden jib, hand-operated crane is used to unload barges at Dickinson's Wharf, Devizes, around the turn of the century. Note the horse-drawn 'hay wain' type wagons being used to transport the sacks of grain.
Boat Museum, Ellesmere

No 148

WILLIAM DICKENSON.
CANAL CARRIER.
BRISTOL & DEVIZES.

Above: **A heavy duty crane is called for to lift this narrowboat onto its 'Crocodile Carrier'. The answer was this pillar crane, make unknown, dating from around 1910.** *Boat Museum, Ellesmere*

Below: **This long reach pedestal crane is capable of reaching the stockpile, the railway trucks and the steam barges from its elevated position.** *Boat Museum, Ellesmere*

Above right: **Priestman crawler cranes on the busy Sampson Wharf in Birmingham in the 1930s accompany what appears to be a somewhat larger Ransomes & Rapier crawler crane. All are engaged in handling vast quantities of steel.** *Boat Museum, Ellesmere*

Right: **High-lift travelling cranes use grabs to handle vast quantities of coal from barges at the GWR Dock in Brentford for the Gas Light Coke Co in the 1950s and 1960s.** *Boat Museum, Ellesmere*

PRIESTMAN
ELECTRA
ETHIOPIA

THE GAS LIGHT & COKE Co
211
690
1056

Above: **Dockside cranes of various makes and sizes set the scene typical at any port in the world where large volumes of cargo are handled.** *A. L. Bailey*

Right: **Among a scene of dereliction is this lonely little hand-operated pedestal crane from the early to mid-19th century. Maker unknown, although it is possibly from Omerod Grierson & Co of Manchester, yet another maker of such cranes in days long gone.** *Boat Museum, Ellesmere*

Below: **This large steam crane made by Thos Smith of Rodley is used to unload railway sleepers for the Hayes Creosoting Depot late in the 1920s.** *Boat Museum, Ellesmere*

Above: **The Universal Crane Co was a division of Thew-Lorain which designed, manufactured, sold and rented cranes. This early truck-mounted crane was, in fact, at the pioneering stages of what has become a multi-billion dollar industry on a worldwide basis — crane hire. This one is using an excavating grab to load that little horse-drawn wagon. Universal became known as the International Cranes Division of Thew-Lorain. The crane featured is from 1919/1920.**
Courtesy PPM Cranes, USA

Centre left: **A P&H Model 200 from 6 June 1924 had moved over from steam power and railway undercarriage to oil engines and crawler tracks. Note the position of the operator — normally he would be facing the work end of the crane.**
Courtesy Norman C. Hargreaves & Mary A. Sellers of PPM Cranes Inc, Conway, USA

Left: **Another truck crane from the International Cranes Division of Thew-Lorain, one of several hundred produced by the company on demand from an ever growing clientele for purchase or hire of mobile cranes in the 1920s.** *Author's collection*

P-H
6915

BUILT BY HARNISCHFEGER CORPORATION MILWAUKEE, WIS.
P&H
P&H GASOLINE DRIVEN
SHOVELS-CRANES
DRAGLINES-TRENCHING
MACHINES-BACKFILLERS
and TRUCK CRANES
HARNISCHFEGER CORP.
Milwaukee, Wis.

Left: **A P&H (Pawling & Harnischfeger) 7½-ton capacity truck crane handling scrap rail lines in Philadelphia for the Pennsylvania Railroad on 3 November 1930. It was designated Model 203A.** *Courtesy PPM Cranes, USA*

Below left: **The Neptune Pageant taking place on 7 August 1926, featuring a P&H truck-mounted crane and the 'Neptune Belles' from Milwaukee, Wisconsin, USA.** *Dave Barnaby — Alatas Crane Sales Co*

Right: **A Jones KL15 mobile crane, fitted with a patent farm grab. A number of specialist farm produce handlers were made; this one, from the 1930s, was designed to ease the task of handling manure.** *G. Self*

Left and below left: **Still hard at work in the 1950s (and some lasted throughout the 1960s) after more than 30 years of operation even then, these views show Ruston rail-mounted cranes. They were probably steam powered originally; some were converted to operate through electric motors for work at Victoria Wharf, Plymouth, where they were mainly engaged in loading china clay.** *ECC Ports Ltd, Plymouth*

Below: **A very popular crane for use in factories, foundries and engineers' yards was this little Rapier crane, which was towed from place to place by a small tractor or truck.** *Royal Engineers Library, Chatham*

Top: Jones of Letchworth, Herts, was also a maker of the range of cranes which began with the smallest KL15 mobile crane, the K144 and K166 in self-contained mobile versions and truck-mounted cranes. Other models included tracked cranes and specially adapted versions for use on docksides and ports with extra-high reach and elevated operators' cabs. They were made by the K&L Steelfounders. This truck-mounted crane on duty for the Royal Engineers is one of many such cranes used by them, both during war and in peace. It is said that the first stores to be delivered to the Normandy beaches in 1944 were handled by Jones cranes. Founded in 1915 by two Belgian industrialists, Jacques Dryn and Raoul Lahy, they became part of the huge George Cohen Group in 1928, known globally as the 600 Group of Companies. By 1938 Jones Cranes was amongst the biggest producers of diesel-mechanical cranes in the UK and exported heavily all over the world. *Royal Engineers Library, Chatham*

Above: A Jones KL66 self-contained mobile crane in the company of the Army in the early 1960s. *Royal Engineers Library, Chatham*

Right: Another KL66 Jones crane at work with the Royal Engineers. *REME Library*

Above: A slightly different truck-mounted crane from Jones Cranes — a Model KL10-10, equipped here with a 30ft jib. It was operated in a somewhat similar fashion to the Winget T-6 and others, from within the cab of the truck; made early to mid-1960s. *Royal Engineers Library, Chatham*

Right: The Taylor Jumbo, complete with a fully hydraulic grab. These, like the Neal Pelican, were popular among coal dealers, although in this case, once again, it is the Army who have found a use for it. *Royal Engineers Library, Chatham*

Below: Further to the Pelican loader, based on a Fordson tractor in reverse, this mobile crane, which can double as a forklift truck, is also based on a Fordson tractor. The maker of this crane is unidentified but companies such as Hyster of the USA would certainly have made similar lifting devices. *Royal Engineers Library, Chatham*

R. H. NEAL & CO. LTD.

Ealing, W.5 & Grantham

CERTIFICATE OF TEST

Cert. No. 5306

Model 'NMS' Crane Serial Number 3315
wheel mounted, and fitted with a 30 ft. Lattice jib
and a LISTER FR2 running Engine, number 20404FR2R6
at a speed of 1250 r.p.m., and tested under the following conditions.

HOIST AND SLEWING TEST

RADIUS OF HOOK	SAFE WORKING LOAD	TEST LOAD	SLEWING SPEED
Maximum 30 ft.	10 cwts.	12½ cwts.	1 Rev. in 15 secs.
Minimum 10 ft. 6 in.	60 cwts.	75 cwts.	1 Rev. in 15 secs.

DERRICKING TEST WITH 12½ cwts. TEST LOAD. TIME MAX. TO MIN. RAD. 18 secs. MIN. TO MAX. RAD. 18 secs.

Travelling Test, with Maximum load on hard level ground, Low Gear 1 miles per hour. High Gear 3 miles per hour.

Gradient Test. The above machine carrying a maximum Working Load at 10ft 6in. Radius of Hook, negotiated the following course 25 yards Gradient 1 in 15. 6 yards level, 14 yards Gradient 1 in 14, 10 yards Gradient 1 in 10.

Stability Test. The Maximum Working Load was derricked out until the wheels furthest from the load began to leave the ground, at this point the radius measured 15ft 3in. This is equivalent to a margin of Stability* of 66%

*The margin of stability shall be taken as the percentage additional load required to bring the Crane to the point of tipping when working on hard level ground.

Other Particulars: Fitted with auxiliary hoist for double rope grab.

I certify that the above machinery, together with its accessory gear was tested by a competent person, that a careful examination of the said machinery and gear, by a competent person after the test, showed that it had withstood the test load without injury or permanent deformation, and that the safe working load of the said machinery is as above.

Signature [signature]
Qualification Chief Engineer
for and on behalf of R. H. NEAL & Co. Ltd.
DATE OF TEST 6.3.
DATE OF ISSUE 1.4.

MACHINE COMPLETED
CERTIFICATE ISSUED TO :- Messrs. Renwick, Wilton, & Dobson, Ltd., 55, Fleet Stree[t] TORQUAY.

ACCOMPANYING THIS CERTIFICATE ARE THE FOLLOWING TEST CERTIFICATES
ENGINE CERT. No. 66541/53813 (Hook V.6054)
HOOK CERT. No. 1/11357/Lot 9938 Ref. 2/38674/35
ROPES CERT. No. R.2339 Ref. 2/40159/6
1/2579/Lot 11777 Ref. 2/47100/12
1/4551/R.20795 " 2/40180/3

Left: All cranes of whatever make or size have to undergo thorough safety checks. Here is a copy of a test certificate made out in 1957 for a Model NMS crane made by R. H. Neal & Co.

SALFORD
HOLYHEAD
LMS
VENTILATED
58 X
LMS

4135

Above left: **The Rapier shop truck crane was unique in that it employed a telescopic boom, long before the modern day hydraulic telescopics. This Rapier crane was made during the early 1930s and was widely used on railway stations, in goods yards and in factories, where its ability to lift and to carry made it an all-round very busy little machine. Available in petrol or diesel, it had a lifting capacity of 1 ton.** *Gordon R. Fairbairn*

Left: **Outside the factory or warehouse we can see the telescopic boom at work when loading this 1930s lorry with huge wooden packing cases. The operator has to stand on a platform at the rear of the crane.** *Gordon R. Fairbairn*

Above: **This little Rapier mobile truck crane of 1-ton capacity has an extended platform.** *Gordon R. Fairbairn*

Right: **The Rapier 'Standard' mobile cranes produced by Ransomes & Rapier from the early 1930s were powered by a combination of petrol and electric, using Ford mass-produced engines and Bull electric motors made in Ipswich. Ironically not only were these wheel-mounted yard cranes used as cranes, but their cantilever type jibs enabled them also to make good use of face-shovel loading equipment which could be used to load coal, sand, gravel and other relatively light materials. These can be seen in illustrations in the section of the book. This crane, one of the earliest of its type, is helping to erect street lights in the very early 1930s.** *Gordon R. Fairbairn*

Above and above right: Another unique feature of the Rapier Standard mobile crane is this special device being used on this 2-ton model, to encompass a drop-ball for demolition work, or in this case at one of the quarries owned and operated by Croft Granite Ltd in Leicester. As a crane the machine would be used to assist with the repair and maintenance of the quarry plant, and to load the granite sets onto lorries. In the two pictures, we see first the crane with the heavy drop-ball firmly in place at the top of the jib, then after the ball has been released to crush or break one of the huge granite boulders. *Gordon R. Fairbairn*

Left: **Is it a crane or is it a shovel? On this Rapier mobile it can perform both duties. One distinct advantage over many other cranes is its ability to align itself to its load, making it very stable under normal conditions.**
Gordon R. Fairbairn

Right: **Another way of handling materials from a stockpile for the Rapier mobile crane, besides using the shovel attachment, was to use a clamshell grab. The machine's ability to slew made it ideal for such work.**
Gordon R. Fairbairn

Below: **Just car parts or whole cars, it made no difference to the Rapier mobile crane. The dockside cranes with a capacity of 3 tons at a radius of 56ft will perform the task of loading the ship, once the goods have been brought within their reach.**
Gordon R. Fairbairn

SPARROWS
CRANE HIRE
BATH
LORAIN
MGL 370

PORT
WILMINGTON

Far left: Many of the 6-ton capacity, lorry-mounted cranes from Ransomes & Rapier of Ipswich made in the 1930s were ordered by the Great Western Railway. Mounted on Thornycroft lorries and powered by Thornycroft petrol engines, they had road speeds of 20mph and capacities were a choice of the 3-ton or 6-ton models. *Gordon R. Fairbairn*

Left: A far cry from those 1920s truck cranes from Thew-Lorain is this Model MC (Moto-Crane) 750, in 1965. *Sparrows/Dave Barnaby/Alatas*

Below left: Over to the Port of Wilmington in the United States to see a P&H 9125 truck crane, complete with elevated operator's cabin,handling ships cargo. *Dave Barnaby/Alatas*

Above: A Thew-Lorain MC-0550 working at the Salt River Project in Arizona, USA in 1965. *Dave Barnaby/Alatas*

Left: In 1963, G. Stiefel Transport AG used a Thew-Lorain MC-9115 Moto-Tower crane to lift this scrubber chimney high onto this plant. A similar crane was used in the late 1960s to hoist a very tall scrubber stack onto a china clay building in Devon. *Dave Barnaby/Alatas*

Right: The RT (rough terrain) range of mobile cranes from Grove starts with the 14-ton Model RT58 to the RT9100 of 100 tons capacity. *Grove International*

Below: The Model TM1500 from Grove is a 150-ton capacity truck-mounted fully hydraulic crane with a standard boom length of 173ft (527m) and maximum tip height of 270ft (82m); a heavy duty crane by any standards. *Grove International*

Bottom: A Grove Coles AT1400 is a 125-ton capacity all-terrain crane mounted on a five-axle carrier. It has a maximum tip height of 63 metres. *Bridon Ropes*

Top right: The 15-ton capacity Grove AP (All Purpose) 415 mobile crane, one of the latest cranes to be added to Ainscough Crane Hire's fleet.
Ainscough Crane Hire/Martin Ainscough

Top far right: A 20-ton capacity Kato, ready for work in 1996/1997 for the hire fleet of Ainscough Crane Hire.
Martin Ainscough/Ainscough Crane Hire

Centre right: An AC 1600 Demag truck-mounted crane at work in 1996, loading this large vessel into the hold of a ship. The crane's capacity is 500 tons.
Ainscough Crane Hire/Martin Ainscough

Centre far right: Working on this new brewery is thirsty work for all concerned with this construction project. The truck cranes of the 1990s are a world apart from their ancestors of the early 1920s. *Bridon Ropes*

Left: The coming together of two very large and successful companies from countries thousands of miles apart is not an uncommon practice. The Japanese company Tadano (formed in 1948, and building its first hydraulic crane in 1955) bought out the interests of the German firm Faun, whose origins date back to 1845. Faun was a fabricator of vehicles, in the past 50 or more years constructing a wide range of very heavy duty lorries, dump trucks, wheel loaders and hydraulic cranes. Now Tadano-Faun truck, all-terrain and rough-terrain, all-hydraulic mobile cranes are being sold, often as complete fleet replacements. One massive order was placed with the company by Greystone, White, Sparrow (GWS) one of the UK's largest crane rental companies. Here is one of their new cranes in their new black livery.
Greystone, White, Sparrow

Top: **Faun all-wheel steer, hydraulic mobile crane in military colours.** *Tadano-Faun*

Above: **A Tadano TG-500E truck-mounted hydraulic crane on a Mitsubishi carrier.** *Tadano-Faun*

Left: **The Faun ATF120-5 five-axle carrier beneath this Faun hydraulic crane.** *Tadano-Faun*

Above: **A Chaseside 1-ton mobile crane from 1954.** *Mr Cutting*

Right: **Some rather unusual types of lifting device from the earlier days of cranage are demonstrated by this Garwood tractor-towed crane, which uses the tractor's cable control unit (CCU) to hoist and lower the load c1930s.** *REME Library, Chatham*

Below: **A 4-ton tractor crane fitted to an early Caterpillar D8 tractor during World War 2.** *Royal Engineers Library, Chatham*

Left: **The base machine of a Ruston-Bucyrus excavator/crane looks something like this when parted from its cab, front and equipment.** *John Laing/Jim Thorpe*

Bottom left: **A 19R-B on dragline duties for the Royal Engineers.** *REME Library, Chatham*

Above: **The NCK was made under licence of the US Koehring Co. Here however is one of the original Koehring machines, one using the excavating grab.** *Royal Engineers Library, Chatham*

Left: **A 10R-B crawler excavator/crane from the early 1950s on dragline duties.** *Royal Engineers Library, Chatham*

Below: **A 33R-B tracked crane stacking HGB decking during the early 1950s.** *Royal Engineers Library, Chatham*

VIRGINIA CONST. CO.
ALEXANDRIA, VA.

Left: **The Osgood company of Marion, Ohio, USA was one of the pioneers in the manufacture of steam powered excavators from the mid-19th century. Most of the early products were steam shovels and draglines. In line with other excavator manufacturers, cranes were to become an important part of their business, not least when their own Universal excavator was born. Osgood continued to manufacture excavators and cranes until 1949, when they merged with the General Excavator Co, also from Marion, Ohio. In 1954 Osgood-General was absorbed by the company which bore the name of the town, the Marion Power Shovel Co. In 1961 Marion also acquired the Quickway Crane & Shovel Co from Denver, Colorado. Marion continued to produce contractor-size cranes and excavators until the mid-1960s, when they concentrated entirely on large mining-size walking draglines, shovels and blast-hole drills. Few Osgood cranes or excavators were made toward the end of the 1950s, by which time production had all but ceased. Here is a little Osgood dubbed the 'Invader' by its owners, a construction company in Virginia, USA, who are using it to top a batching plant using a grab.** *Peter Gilewica, Marion PS Co*

Below left: **This large electrically powered Marion Model 480-E is working on a big pipeline project alongside a host of other crawler cranes in September 1936 for its owners, Morrison & Knudson, Riverside, California.**
Peter Gilewica, Marion PS Co

Above right: **The Marion Model 42-M was one of the most popular models in the Marion contractor-size range of cranes.** *Peter Gilewica, Marion PS Co*

Right: **Working for its owners, Leonard Brothers Transfer, of Orlando and Miami, Florida, is this Model 43-M truck crane from Marion.** *Peter Gilewica, Marion PS Co*

MARION
FOUNDATION CO.
NEW YORK
MARION
362
MARION

Left: This Marion 362-M on grab duties will no doubt be invited to stay on to assist with the rebuild on this site, once initial demolition has been completed. The site is in downtown New York. *Peter Gilewica, Marion PS Co*

Above: A Priestman Lion equipped with a 4-ton drop hammer operated from a crane winch, driving wooden piles for sea defence work. The normal crane jib has been dispensed with and the leaders (hammer and guide structure) are close-coupled to the crane, thus improving stability and allowing a potentially longer leader to be used. The verticality of the leader is adjusted by hydraulic cylinders. The site is on the East Anglia coast, near to Lowestoft. *BSP International Foundations Ltd*

Above right: One of the most popular cranes throughout the 1950s and 1960s was this, the 'Smith 21' from Thos Smith of Rodley. These were, in fact, built for Smiths by Butters Bros of Glasgow. At one period, Butters was best remembered for its range of 'Scotch Derricks'. This Smith 21 is using a double rope, whole tine grab to dredge this river, working from a pontoon. *Author's collection*

Right: A winch-operated drop hammer with the leaders attached to the top of the NCK-Rapier crane jib. Interesting to see a wheel-mounted crane from R. H. Neal being used on this site in the 1960s. *Roger Elliott, BSP International Foundations*

8OCA
'terradrill'
'terradrill'

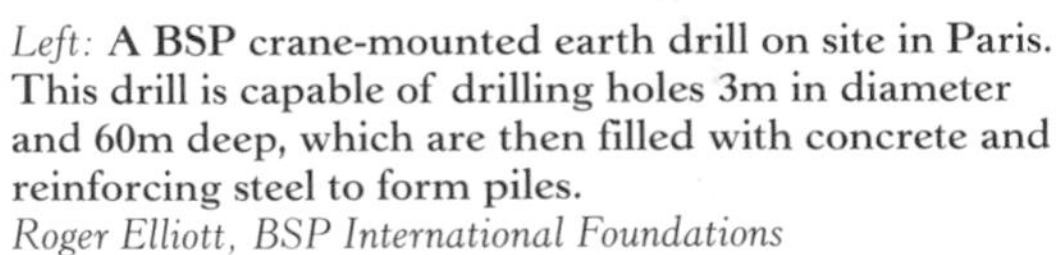

Left: A BSP crane-mounted earth drill on site in Paris. This drill is capable of drilling holes 3m in diameter and 60m deep, which are then filled with concrete and reinforcing steel to form piles.
Roger Elliott, BSP International Foundations

Above: In 1967 this Priestman Bison BC72, a 30-ton capacity crawler crane, had been fitted with a tower crane attachment. *Mr Peacock*

Top right: One of the heavy duty crawler cranes from P&H (Harnischfeger), using its fly jib to ever-so-carefully position parts at this petro-chemical plant in the late 1960s/early 1970s. *Norman Hargreaves*

Centre right and right: In this pair of photographs we see two of the latter generation of Priestman Lion cranes (on dragline duties) made just prior to the closure of Priestmans, who had been manufacturing excavators from the middle of the 19th century. The first shows the Lion 40H loading a 6-wheel dump truck. The second shows the Lion 50H throwing out its dragline bucket ('casting', as it is known).
Mr Peacock

Left: **Two Priestman Tigers equipped as draglines are hard at work in an opencast mining site in South Wales during the early 1960s.** *Short Bros*

Right: **A 19R-B from the early 1960s equipped with a grab bucket is ready for transit to its contract on a coal site in South Wales.** *Short Bros*

Below: **One of the new generation of hydraulically operated crawler cranes from the NCK factory in Ipswich is this special dockside version, with an ultra-elevated operator's cab enabling the driver to see clearly into the holds of ships.** *NCK*

Left: The Link-Belt Model LS-80 is one of the new generation of fully hydraulic crawler cranes, undergoing acceptance testing at the LBS (Link Belt Speeder) SPA factory in Italy. Link-Belt can trace its origins back to when the first excavators and cranes were produced at its factory in the United States. Since then, Link-Belt machines have been manufactured in Chicago, Illinois, Japan, Brazil, Mexico and Italy. In 1894 Link-Belt built rail-mounted cranes capable of handling coal using a clamshell (grab) and locomotive cranes. In 1939 Link-Belt purchased the Speeder Machinery Co which was manufacturing light excavators. In 1948 the manufacturing facilities for the Link-Belt Speeder Co were switched to Cedar Rapids, Iowa, USA. *Link-Belt, Italy*

Above: The NCK Nova HCV50 is one of the totally new all-hydraulic cranes from the company which started building machines with this logo in Thornecliffe, near Sheffield in 1947. Prior to that they were known as NCH (producing excavators and cranes under licensing agreements with the Harnischfeger Corporation of the USA, makers of the P&H range). NCK grew from a licensing deal with another US manufacturer, Koehring. (NC were the initials of George Newton and Thomas Chambers, who founded the original company in the late 18th century.) In 1958, Ransomes & Rapier was added to the NCK company. By the 1970s the Rapier partnership had ceased. The Nova HC50 here is carrying a diaphragm wall grab. *NCK*

Above left: **Sporting a very long main boom and a fly jib is this new NCK Nova HC65 all-hydraulic crawler crane.** *NCK*

Above: **From Mannesmann Demag Gottwald of Düsseldorf, Germany, we see the very latest high-tech mobile harbour type cranes, with the model HMK90H loading an articulated road haulier at Kelheim, Germany in the mid-1990s.**
Mannesmann, Demag Gottwald

Left: **At Napier, New Zealand, a Gottwald HMK280E is loading the cargo vessel *Forum New Zealand II*. Note the Hitachi hydraulic excavator on the deck of the ship.**
Mannesmann, Demag Gottwald

WALTER WRIGHT

Left: When it comes to heavyweight crawler cranes, few can match the truly massive spectacle of those made by Neil F. Lampson of Kennewick, Washington, USA, whose Trans-Lift mobile heavy lifting systems use two diesel engines to power crawler transporters. These support the main boom and mast on a forward transporter, which is structurally connected to a second transporter which enables the device to steer. It carries the counterweights and strut to the top of the mast. The main boom is normally 122m in length and is able to use a jib of 61m with a mast of 59m or 67m. The crane is capable of lifting 1,815 tons and is able to travel with such heavy loads. The model has been produced since the early 1980s. The machine featured here was at work in the off-shore oil industry near Aberdeen. *Bridon Ropes, Doncaster/DGS Films*

Below left: Some of the real heavyweights in the crawler crane industry are taking part in big lifts on this construction site. The large crane in the foreground is made by American, whose origins date back to 1882, the present name being adopted 10 years later. In those days the company manufactured hand winches, wooden derricks and steam locomotive cranes. Now with over 100 years of experience to draw from, they are producing some of the world's largest crawler cranes, some of which come equipped with a 'sky horse' attachment to enable them to double the lifting capacity of the crane without outside help from other cranes. These machines are currently reaching capacities approaching 1,000 tons. The cranes in the background were made in Japan by Sumitomo Construction Machinery. These have lifting capacities in excess of 250 tons for the top-of-the-range models. *Bridon Ropes/Neil Richardson*

Above right: In 1981 at the Esso refinery in Fawley, these cranes engaged in a multiple lift which was considered one of the heaviest in the world at that time — obviously several thousand tons. Multiple lifts are a common occurrence, particularly in the off-shore oil industry, where huge platforms have to be constructed and turned on their side and eventually lifted into a position for towing out to sea. Lifts as heavy as 13,000 tons have been undertaken using as many as 20 huge American and Manitowoc crawler cranes, with occasional help from the huge Lampson. *Bridon Ropes, Doncaster/Neil Richardson*

Right: An NCK-Rapier heavy duty crawler crane hitches a lift down into the workplace, on the construction of the new Jubilee Line in London. Big Ben is well placed to watch the proceedings. Evidently one large crawler crane had already made the journey. *QA Photos/Burne Craigie Photos*

Left: This heavyweight Gottwald crane is lifting a 420-ton reactor with no additional help. *Bridon Ropes/Neil Richardson*

Below left: A massive Demag truck crane, a crawler crane, one other truck crane, numerous forklift trucks and mechanical handlers, along with one little mobile crane made by Henry Gooch and that 30-ton capacity overhead gantry type crane by Pellaby, set the scene on this central London construction site. *Balfour Beatty/Byrne Craigie/QA*

Right: The North Sea oil and gas industries have been major users of some of the largest crawler cranes since the 1960s. Throughout this period, the vast majority of large cranes have been Manitowocs, American Hoist & Derrick and the huge Lampson, with backup cranes from NCK-Rapier and top-of-the-range Ruston-Bucyrus heavy duty models. The BP Magnus jacket is seen being floated out from Highland Fabricators Ltd's Nigg Yard in Ross-shire, one of the biggest sites on which heavy duty crawler cranes operate. Here we see the Lampson being assisted by two Manitowocs and a hydraulic mobile crane on the near bank, while at least one other Manitowoc is just visible on the far bank; the tug boats ever-so-carefully tow the huge structure out into the open sea. *DGS Films*

Below right: The Lampson 1,350-ton crawler crane nearest the camera, with the other crawler cranes on duty nearby. The cranes are lifting a mud-mat while construction of the jacket proceeds. *DGS Films*

Below far right: A Lima 2400 is much more likely to be seen equipped as either a face-shovel or as a dragline. On this occasion however it is working as a grab-crane from a barge, on river cleaning duties. *Bridon Ropes*

Right: Some recent innovations in crane uses include this hydraulic crane from Japan, mounted on a tracked runabout. It features a 1-ton capacity Tadano telescopic crane on a Kubota RC25P tracked carrier. *Chapple Plant Sales*

Below: Cranes have to be tested on a regular basis, as do all forms of lifting tackle, even if only a set of chains or a set of ropes. This does prevent many accidents; however, accidents still do occur all too often with all types of mobile, crawler and tower cranes. A very large NCK-Andes C75 crawler crane has crashed while negotiating a ramp which was evidently too narrow and had been built with unstable material. One of the two cranes in the vicinity is using a power-driven drill, while the yellow crane will no doubt be invited to assist the Andes back onto its tracks. In the immediate foreground three Poclain hydraulic excavators are engaged in earthworks on the site. *Tony Swaine*

Left: The NCK-Andes is rated at a little over 30 tons, but this tank, weighing 45 tons when the contents were taken into account, has resulted in a snapped boom; just another example of accidents involving cranes. *Tony Swaine*

COMPANY HISTORIES

Case studies of a few famous manufacturers

Stothert & Pitt

From 1844, the company name Stothert, Rayno & Pitt was well known as a manufacturer of such items as cast iron oven doors. The involvement of Henry Stothert, one of the founding fathers of the company, in the manufacture of railway engines, steam ships and horizontal compound engines, popular at the time with the country's textile mills and, occasionally, bridges, were an example of the extent of the early engineering involvement of the company.

Stothert were known to have built early cranes during the 1850s. These were hand operated. A three-ton capacity version was installed at Carmarthen in 1850 having a steel jib. Other early wood jib cranes were used at Box Quarries from around 1855 to 1883.

Crane making began to dominate the firm's activities from the 1860s. An order for a steam powered crane for Bath Stone Quarries was received in 1853, although this by no means signifies it as being the first steam crane to be manufactured by Stothert & Pitt.

Above: **The cargo vessel *Benloyd* discharging at Liverpool. On board are Stothert & Pitt 5-ton capacity electric cranes.** *Bridon Ropes*

Left: **A trio of Stothert & Pitt electric cranes at the side of the Birmingham Canal at the Tyseley Goods Yard, loading steel onto the narrowboat *Tiverton*.** *Boat Museum, Ellesmere*

Neither were they the first recorded makers of steam cranes, as in 1839 James Taylor of Birkenhead made just such a device for use in a stone quarry. Similarly the same man built the first travelling crane powered by steam to a design presented by W. W. Hulse. Finally completed in 1858, a steam powered overhead travelling crane, made also to the design of Hulse, was in operation by 1852.

Robert Hawthorn, however, made a steam travelling crane as early as 1824. Other companies making steam travelling cranes in 1865 or 1866 were Appleby Brothers and Cowans, Sheldon & Co.

Certainly in 1865, a Stothert & Pitt steam crane, which travelled along railway lines, was in use at the Port of Swansea, while a year earlier a similar crane is believed to have been in use in Guernsey.

Capacities of three tons and six tons were being offered. Indeed a six-ton crane was on show at the Paris Universal Exhibition of 1867, at which it won a Silver Medal. Five-ton capacity cranes were to win Gold Medals at exhibitions in London in 1885 and again in Paris in 1889.

In 1883, a patent was granted to Joseph Henry Wild of Leeds, for a grab which could handle mud, sand and grain. The grab used a single chain to operate the opening, closing and vertical movements and was operated by one man. This device was ably demonstrated on the Stothert & Pitt steam crane in Paris in 1867.

In 1876, Stothert & Pitt built steam cranes to the patent of William Fairbairn of Manchester. Fairbairn also had manufacturing facilities for his patent cranes, six of which he made for Keyham Dockyard near Devonport. These had the advantages of riveting together the iron plates of the projecting arm to form a hollow tubular girder of curved shape to replace the normal straight wood or iron jib. He took out a patent to protect this idea in 1850. It allowed bulky articles to be raised much nearer the jib sheave than would normally be possible. It was an ideal tool for a dockside crane. Such machines are still in existence in Bristol, and another unidentified make at Dover.

Stothert & Pitt was to become well known for the manufacture of huge cranes known as Mammoth, Hercules or Titan. They were suggested to Stothert & Pitt by Consultant Engineer William Parks and Resident Engineer W. H. Price, both involved in the construction of a harbour at Karachi (now in Pakistan).

These huge cranes were able to lift huge blocks of stone or precast concrete. The Titan was capable of lifting and placing blocks weighing 27 tons. A Titan weighing 576 tons was made for use at Peterhead Harbour of Refuge in 1891. It was capable of lifting 50 tons at a radius of 100ft. A very large crane named the Goliath was used in conjunction with the Titan.

Electrically operated cranes were first used around 1887 in Hamburg. British electric cranes were built by Mather & Platt of Manchester, while another was designed by W. Anderson. Stothert & Pitt received an order in June 1892 for two electric travelling jib cranes of three-ton capacity, destined for Southampton Harbour.

Through the years, with orders from all over the

Above: **Stothert & Pitt dockside cranes are at work in ports all over the world. These examples are in the Middle East.** *Stothert & Pitt*

Right: **This oilfield crane was made by Stothert & Pitt of Bath and Bristol in December 1978.**
Bridon Ropes/Stothert & Pitt/Clarke Chapman Marine

world for taller, bigger, more efficient cranes, Stothert & Pitt would grow to be unmatched, especially where dock cranes were concerned. The cranes have now changed to keep pace with modern methods of handling vast quantities of freight, now almost all of which is containerised.

Stothert & Pitt is now a member of the huge Clarke Chapman group, itself a component of Rolls-Royce. Among the companies sharing this unique umbrella are Cowans Boyd and Wellman Booth, who in common with Clarke Chapman itself were formidable crane makers. Today, along with the Stothert & Pitt arm of the company, they are well placed to continue producing the world's cranes well into the next century. That is, however, a very bold statement to make in an age when technological progress can overthrow any predictions.

Coles Cranes

(Henry Coles 1847-1905)

Charles James Appleby of Appleby Brothers of Southwark, London, made the firm a pioneer in the manufacture of steam travelling cranes, with particular emphasis on those used exclusively as railway breakdown cranes.

Many of Appleby's former engineer employees left to create their own crane-making businesses. Alexander Grafton was one, Henry Coles another. Appleby Brothers moved from their original factories, one at Emmerson Street, Southwark, the other at Sumner Street (which was merely a small workshop), to new premises at East Greenwich.

This was the opportunity Henry Coles was looking for. Joined by his three brothers, Frederick, Walter and Ernest, he moved to the Sumner Street workshop in 1878, to begin what was to become a long and prestigious crane-making business. Henry's brothers had also been employed by Appleby, so they were well placed to take advantage of this new-found opportunity.

By the 1890s, the range of Coles cranes were finding favour around the world. Most were rail mounted, steam powered and were offered with or without a Coles single chain grab. In fact, several fully-slewing rail-mounted steam cranes were sold to Glasgow Corporation Gas Works in 1895. All were

Above right: **A Coles EMA (Electric Mobile Aerodrome) crane, lifting a boat for the Royal Engineers in the 1930s or early 1940s. Both the British Army and the Air Ministry placed orders for Coles EMA cranes in ever increasing numbers from 1930 onwards, with a request for 82 2-ton capacity cranes being received and a further 120 ordered in 1930. Once taken over by the Steel group of companies, whose works were in Sunderland, much of the manufacturing capacity for Coles Cranes was transferred to the Northeast, and because of the many orders from the various ministries, it was decided to name the factory Crown Works. It is still a major player in the production of cranes, now under the banner of the giant American manufacturer, Grove, who had acquired the rights to build Coles Cranes, Allen of Oxford and their own line of mobile, crawler and truck cranes.** *Royal Engineers Library, Chatham*

Right: **A front view of a Coles truck-mounted crane on its way to work for the Army.** *Royal Engineers Library, Chatham*

Right: **This Coles mobile crane is almost certainly working for the Air Ministry, judging by the aircraft propellers in the background. They were very useful cranes to have in an aircraft factory, with the cantilever jib and ability to operate in areas with low headroom.** *Royal Engineers Library, Chatham*

Below: **Yet another Coles EMA cantilever boom truck crane from the 1930s.** *Royal Engineers Library, Chatham*

fitted with the grab which earned Coles recognition at an International Inventors Exhibition in 1884. A similar crane using a hook for general crane duties was supplied during the 1880s for work at Jersey Harbour.

Coles' cranes and excavators had become so innovative that by 1895 the company catalogue was able to list steam powered cranes, electrically powered cranes and cranes powered by hydraulics. That is to say water driven, taking their power from water standpipes at the side of the track. This was not a totally new concept as hydraulic cranes had been in use from as early as 1847, some being installed at Newcastle Docks in that year.

A Goliath crane, powered by steam, was built for harbour construction on one of the Greek islands. Unlike the earlier mentioned Stothert & Pitt Titans, this one took the form of an Overhead Travelling Crane. It appears the Goliath of 1887 was the only such machine sold, although similar machines were offered in one of the Coles catalogues in 1894 for £550. Other overhead cranes, fixed based cranes and excavators were certainly on offer, along with the company's very successful range of fully-slewing steam cranes.

The excavators referred to were devised by French engineers, Couvreus and Bourdon, for use on the Suez and Panama Canals. These were endless bucket chain excavators, used to cut excavations and make embankments. Another excavator being manufactured by Coles, was the Gatmell, devised by a Mr Gatmell for sinking cylinders on the Empress Bridge over the River Sutlej.

However, by the end of the 19th century, much of the major canal building contracts had been completed, so the requirement for such machines was more or less over. Many large German manufacturers of such equipment were able to cope with the existing demand, mainly in the mining of lignite and quarrying cement ingredients (chalk and clays), as well as harvesting potato crops.

By 1898 Coles had outgrown his factory in Sumner Street and so acquired a site in Derby for an all-new factory, which by comparison with the earlier premises was huge. He continued to call it The London Crane Works. This move allowed for greater expansion. From this move, Coles was able to bring about the development of a whole new generation of rail cranes, which included the 1907 40-ton rail crane, along with specialist cranes for use in steel foundries.

In the early 1920s work began on designing a totally mobile crane which operated off the back of what had been, certainly on the prototype, a bus chassis. The eventual crane would be the Tilling-Stevens petro-electric solid-tyred bus chassis, supporting a special single-motor superstructure. Such machines were to be much sought after, not least by the military on both sides of the Atlantic, to assist with the lifting of heavy guns, ammunition and supplies. It was also sold in large numbers to customers in Japan, the Karachi Ports Trust and users in the UK.

Henry Coles died in 1905, two years before the company's 40-ton machine was finding favour in steelworks and ship-building yards, particularly in the northeast of England, and a full three years after the company had experimented with an electrically powered crane, run on similar lines to that of a trolleybus, not the most successful experiment in the company's history!

His brothers, Walter Joseph, Harold Lewis and Ernest Coles, and his eldest son, Henry, carried the business forward. Henry was killed in action in France in World War 1, the brothers all dying during the 1920s. The widows of the Coles brothers, whose families were entirely daughters, decided to sell the company and so from October 1926 it was, for the first time since its inception, run by people who did not bear the name Coles. They were William Searle, William Robinson and Alfred W. Farnsworth.

The day-to-day running of the business was left to Arnold Hallsworth, who had earlier been entrusted to design the EMA mobile crane by Coles. With the looming depression of the late 1920s and certainly by 1930-1, orders were decidedly thin on the ground. From 1930-5, orders for just two or three cranes a year were far from unusual. Only by its ability to supply vast quantities of spare parts for machines previously in service was the company able to survive. Fortunately, better years followed, although it was the outbreak of war which saw an unprecedented request for cranes, not least by the Air Ministry, who in one go placed an order for 82 two-ton capacity cranes on 13 July 1937. This represented the largest single crane order ever placed with a British company. The original order, which was later reduced, had been for 120 cranes!

Two cranes had been supplied mounted on Morris chassis with full pneumatic tyres in 1936, and this at least gave some indication as to what would be required. Other British manufacturers had relied heavily on a three wheel 'castor' chassis for the slewing arrangement. This was an alien design for Coles, so despite such specifications coming from the Ministry, Coles proceeded to design and build cranes of a type they were much more used to, being lorry mounted or at least having their own wheel-mounted, independent drive chassis, on which a fully slewing crane was mounted. As a consequence, the Coles design was favoured and adopted by the powers to be at the Air Ministry.

During those following years, the company order books were full. In 1939, 60 cranes were built at the Derby plant, while 120 mobiles were ordered in 1939, by which time the Henry J. Coles company had been acquired by the Sunderland-based Steel & Co, where many, although not all, of the further orders for Coles cranes would be despatched.

As World War 2 came to an end, the manufacture of all Coles cranes from the late 1940s until the 1960s was transferred to the Northeast. By 1954, Coles was producing a very wide range of rail, truck and self-

Below: **A Coles mobile crane seen working on an airfield in the 1950s; just one of a very wide range of options from Coles Cranes catalogue in the 1950s and 1960s.** *Royal Engineers Library, Chatham*

Left and below left: **Further examples of Coles cranes at work. All are truck mounted and all are working for the Royal Engineers.** *Royal Engineers Library, Chatham*

contained mobile cranes, featuring either a cantilever type boom or a traditional lattice boom mounted at the bottom of the slewing superstructure, while experiments were already under way with fully telescoping booms. Then came Colossus, then the world's largest mobile crane. With a capacity of 41 tons, it was truly the largest truck-mounted crane available in 1954. In 1963, another first for Coles was the introduction of a truly mammoth truck crane, the Coles Centurion, with a capacity of 100 tons. By 1971, the 200-ton barrier was breached by the Coles Colossus 6000.

Diesel-electric drive cranes had been the mainstay of Coles products for over 25 years but by the late 1950s other options were being considered by Coles, who could see certain competitors bringing into use diesel-mechanical drives through much improved transmission systems, and the increased use of hydraulics was already on the horizon.

To keep pace with or ahead of current thinking, Coles looked to acquire companies such as R. H. Neal & Co of Grantham, whose use of diesel-mechanical drive systems were attracting large orders for their cranes, while F. Taylor & Sons of Manchester had near perfected the totally hydraulic crane. Both companies became part of the Steel Group of companies in 1959.

R. H. Neal had been manufacturing a range of cranes from the 15-cwt capacity GM Mobile, which used a 15-ft jib, and the very successful Neal NS46 of 4-ton capacity, both extremely successful throughout the 1950s.

The Pelican Hydraulic Loader was a hydraulic grab, mounted on a hydraulically raised boom situated at what looked like the wrong end of a Fordson tractor chassis. In fact, the tractor had been reversed, in that the engine and steering wheels were behind the driver, while the large driving wheels and loader were now in front. This was the perfect tool for loading or unloading material from railway wagons, such as coal for example. Many such machines would subsequently be found in coal handling yards around the world. The original design and manufacturing licence for the Pelican had been acquired from a New Zealand manufacturer.

Neal also manufactured other products under licensing agreements, including the Hymax hydraulic mobile crane and from the USA they made the Unit cable excavator/crane, known here as the Neal Unit. It was certainly one of the first British-built universal excavators to feature a cab for the operator which was outside the confines of the main machinery house. Priestman was to follow this idea with the Cub and the Lion. On many of the cranes by Coles, Jones, Neal and others, it had long been a common practice to separate the operator from the noise of the engine and main machinery, something which is much more common today on both cranes and excavators, whether hydraulic or cable types. R. H. Neal had in fact been making cranes since 1930.

When F. Taylor joined Coles at the Steel Group in 1959, it had already been making hydraulic cranes for more than 10 years. Its first model was known as the Coffin. It was not able to slew, it could not raise or lower its jib; it was, however, useful as a yard crane.

Above right: **A Coles Aeneas truck lifting steel slabs for Peach & Tozer Ltd at its stockyard in Rotherham. It was a branch company of the United Steel Co. Cradle slings manufactured by British Ropes at its Doncaster works are being used in this lift in March 1964.** *British Ropes, Doncaster, with thanks to Neil Richardson*

Right: **The Coles Husky of 1966/7 was born out of the marriage between Coles Cranes and F. Taylor. Both were able to use each other's experiences to develop what was to become one of the earliest 'all terrain' cranes, featuring a fully telescopic boom. This type of crane is now manufactured by a whole new generation of companies around the world and is a common sight on both our highways and construction sites. This 12-ton capacity Hydra Husky was advanced for its day.** *G. W. Sparrow & Sons*

TWIN BOLSTER
SET 629

COLES
HYDRA HUSKY
SPARROWS
CRANE HIRE
PHONE AVONMOUTH 3000-2970
HEAD OFFICE
BATH
PHONE - 21201
(10 LINES)

Future models, though, were fully slewing, had telescopic booms and were often mounted on a four-wheel drive chassis. The Taylor Series 42, for example, became a firm favourite with the military. It was used by the army in large numbers, by the RAF and was useful on aircraft carriers.

The Taylor Series 50, a general-purpose mobile, was a familiar sight on the roads, running between one lifting contract and the next. It too, was the perfect tool for work in factories, installing or dismantling machinery in storage yards, scrap yards and other sites (timber, steel, iron foundries). The Taylor Jumbo was by far the most famous from this company, being equipped to operate a hydraulic grab. It was popular among coal yards in particular.

The development of the Speedcrane in the 1960s brought to the crane hire business a development which would last for decades. It was a rapid response crane which was able to travel the country's roads with ease, was able to work in relatively confined spaces and had the advantages of a telescopic boom. It was also fully hydraulic. Whether it was a building site, vehicle recovery, factory machinery installation or dismantling, lorry loading or a stacking yard, it was the ideal runabout. Having front wheel drive and rear

Left: **A 25-ton capacity Coles crane from 1962 in the livery of G. W. Sparrow Crane Hire of Bath. It is a Model 27.10.** *G. W. Sparrow & Sons*

Below: **In 1964 when this huge 100-ton capacity Coles Centurion was on the road, it was the largest mobile crane in the world. Once again it was G. W. Sparrow of Bath who chose to purchase it.**
G. W. Sparrow & Sons

wheel steer made it totally manoeuvrable. It became the forerunner of many such mobile cranes, many of which are still among the best-sellers today and are available from manufacturers worldwide.

Coles was to adopt the Taylor concept into a new line of mobile hydraulic cranes under the name of the Coles Hydra, originally a Taylor trade name.

Gordon Innes, CEng, FIMechE, MSAE has provided a personal viewpoint of his involvement with Coles:

'Coles cranes were, in the early days, distinguished by being petrol-electric powered using a variable speed crane. The early Ransome cranes also used electric transmission but ran their engines at constant speed and controlled the output speeds by resistances. The Coles solution was cheaper, more instinctive and more readily developed for larger machines.

'Coles Cranes was taken over by Steels Engineering Products of Sunderland in 1937-8. The owner, Eric Steel, did a clever selling job at the Air Ministry, and Arnold Hallsworth produced the EMA design mounted on a Thornycroft chassis for aircraft maintenance. I believe that the earliest version was the screw-derricked version, with the screw and the hoist motion interconnected by a dog clutch to give an element of 'level luffing', ie the load moved approximately horizontal when the jib was derricked (luffed). The nut on the derricking screw was prone to wear, allowing the jib to drop, until a safety nut was fitted.

'In 1945, with Arnold Hallsworth as MD, they engaged an industrial designer to produce a modern design of crane. This, I believe, was called the 610. The cab was a perspex bubble and looked fabulous, but scratched in service, so panels of glass were then inserted.

'When I joined Steels in 1950, they manufactured Euk industrial catering machinery and farming machinery, Electric Eel battery-electric factory trucks, overhead travelling cranes and battery-electric delivery vehicles, as well as mobile cranes. They also ran the builders' hardware business from which it had all grown under Eric Steel's inspired leadership.

'Outriggers were fitted to the 610 and the structure beefed up to make the 1010, which grew to be the 1110. When the next size step was considered, it was decided that the cantilever jib was not suitable and therefore a 20-ton lattice jib crane was produced. Shortly after its prototyping, a design for a tubular lattice jib from Tubewrights was bought, together with design calculations. I studied these and from then on our own designs were produced, because we felt some of the assumptions made by Tubewrights were invalid. There followed a wide range of truck and mobile cranes, including the bridging crane used to launch Bailey bridges, and high level boom cranes for New York Dockyard, with a degree of level-luffing. Another machine was the huge 4612 crane with a capacity of 45 tons at 12ft radius, based on the running gear of a Thornycroft Antar tank transporter. Coles was the first European mobile crane maker to use very high tensile weldable steel for jibs and for chassis structures.

'As the company grew, Taylor Jumbo Cranes and R. H. Neal of Grantham were taken over. At about this time we were offered the manufacturing rights for 'Hiab' lorry loaders. Reg Keats the Marketing Director told the board that it was an unsaleable idea, so they did not proceed with it. The chief engineer of Taylors designed and built, without board approval, a 12-ton telescopic jib crane, which Keats described as "a flash in the pan". However, I left Coles just before this.'

When Gordon Innes returned to the business in about 1969, he presided over the departure of the Roadmaster. This was based on a commercial chassis which was not designed for sustained travel in reverse, with resulting service unreliability. They developed a design for the 971 HLB, based on the 851 mobile crane and the design they had made for New York Dockyard years before, and Ian McLeod-Smith sold the design 'off the drawing board' to two Swedish ports.

When they took over British Hoist & Cranes, they inherited the Iron Fairy designs. They did a design review and, based upon them, Ron Millyard and Gordon Innes produced the designs for the IF6, IF8 and IF10 in two weeks of frenetic drawing and calculation. The main new feature was a more sophisticated chassis capable of taking bolted-on outriggers.

A few years later, the IF12 and IF15 were produced, the first mobile road crane with automatic transmission. Telescopic jibs at this time all had internal chains, hoses or ropes to drive the sections which, being out of sight, tended to be neglected. Their new machines eliminated these with a patented design.

Clarke Chapman & Co Ltd

The company began life in 1864. Its founder was the engineer William Clarke. The factory was situated at Felling Shore, close to Gateshead on the River Tyne, and manufactured small hand driven winches.

In 1868, Clarke designed and built the first steam driven cargo winch. Donkey boilers were built to provide the steam necessary to power them. Six years later a steam driven windlass was added to the manufacturing product range and the company moved to the Victoria Works site in Gateshead with a workforce of 350 men. Already Clarke Chapman was becoming a respected name in the marine industry and in 1874, Captain Abel Henry Chapman joined the company as a partner.

In 1884 Charles A. Parsons joined the company as

BEN OCEAN LANCER

Above: **This heavy duty crane was made primarily for the off-shore oil industry for work on drilling rigs.** *Clarke Chapman Marine*

Above left: **A deck-mounted cargo crane from Clarke Chapman Marine, ready to handle most types of ships' cargo.** *Clarke Chapman Marine*

Left: **This ship-mounted drilling rig, the *Ben Ocean Lancer*, has two heavy duty cranes aboard to handle the heavy drilling rods and other vital stores and equipment.** *Clarke Chapman Marine*

a junior partner. His arrival coincided with the development of electricity for shipboard use. Park House, Gateshead, which stands alongside the works, had been purchased for this purpose.

The company was known as Clarke, Chapman, Parsons & Co and in the same year constructed the first steam turbine dynamo. This machine became an instant success and marked a revolutionary change in steam engineering history. The output of the first turbo-dynamo was 7.5kW at a speed of 18,000rpm, 100V. The dynamos were mainly used on board ships for electric lighting.

At the 1885 London Inventions Exhibition, the company exhibited a turbine which was the most talked about item at the show. During the same period, Parsons had begun to manufacture vacuum incandescent lamps, ranging from 100 to 3,000 candle power and running on a 100V supply.

By 1888, Clarke Chapman generating, lighting and deck equipment was becoming generally used by the Royal Navy. The partnership between Parsons and Clarke Chapman ended later that year, to enable Parsons to continue development of the turbine. After a great deal of legal wrangling concerning the patents, he moved to Heaton Works, Newcastle, having been forced to modify the design.

In 1891, Edward Butler, a pioneer of the gas and oil engines joined the company as an engineer. Butler had earlier patented a device called a Velocycle which was basically a tricycle driven by a petrol engine. The device was considered too fast for use on public highways, reaching speeds of up to 15mph. Indeed part of his design included the forerunner of the modern carburettor. One of his engines and a dynamo were installed together with full electric lighting at the Theatre Royal, Newcastle. His engines were also in use throughout Victoria Works itself, driving overhead shafting to power the main machine shops.

By 1901, the workforce had risen to over 2,000 and the company had gained an international reputation. They produced the first electrically driven windlass in 1902, although development into new steam driven products continued well into the 1950s. During World War 2, the company was one of the few manufacturers to supply complete searchlight units, including lamp, projector and reflector. Indeed they helped solve major problems for the armed services by producing a more efficient reflector and a lamp that was resistant to shock and marine corrosion.

In a one-year period the company sold over 45,000 winches.

The product base was continually developing. By the 1970s, Clarke Chapman Marine, as it was then known, produced a wide range of marine deck machinery, ship and cargo handling equipment for vessels of all sizes and types, including winches, cranes and capstans. Offshore equipment included anchor windlasses, winches, fairleaders and pedestal jib cranes using the most modern electrical thyristor-controlled ac and dc drives, temperature controllers, analogue and digital control systems. All products could be adapted to both land and marine use. The company also produced a wide range of specialist equipment for both the UK and foreign navies.

The electric cranes manufactured covered the

Above: **John Thompson/Clarke Chapman onboard ship's cranes in March 1973.** *Bridon Ropes*

complete range of cargo handling, bulk material and container applications up to 50 tons. The company had also considerable expertise in producing a range of boilers, from small industrial usage to massive coal-fired and nuclear power station applications.

In 1973, the company had grown into eight main divisions, located throughout the country. These were:

Marine & Engineering Division — Gateshead, Tyne & Wear
Power Plant Division — Gateshead and Wolverhampton
Crane & Bridge Division — Leeds, Carlisle and Annan
Pipework & Pressure Vessels Division — Wolverhampton
Thompson Cochran Division — Annan and Lincoln
Transporter Division — Dudley and Sheffield
Horseley Piggott Division — Durham
Advanced Technology Division — Gateshead

The structure remained thus until the formation of Northern Engineering Industries in 1976, created by the merger of Clarke Chapman and Reyrolle Parsons. Together they formed a new and fully functioning multi-million pound engineering group operating from over 60 locations in Britain.

Manitowoc

In 1902, Charles C. West and Elias Gunnell, who were employed by the Chicago Shipbuilding Co, started looking for a shipyard to buy. West and Gunnell had both become dissatisfied with their employers, especially after they had been absorbed through a merger by the American Shipbuilding Co of Cleveland, Ohio, in 1899. As a result of the merger, Chicago had lost its identity as one of the most progressive shipbuilding companies. Instead it became merely a loan subsidiary, with no real decision-making power.

Charles C. West was a marine engineer and naval architect while Gunnell was an experienced shipbuilder, designer and mechanic. Both had the visions of making steel ships rather than the wooden ones being fashioned by other companies. However, they received much advice from other shipbuilders, part of which was to continue working with wood, certainly in the ship repair context, as this would have enabled them to accumulate enough money to sink into their eventual aim, that of building steel ships at the yard they had located, which was for sale. This was the Burger & Burger yard in Wisconsin on Lake Michigan.

This yard, through its retiring owners, Henry and George Burger, had built a good reputation for quality craftsmanship, again while working on wooden boats. Other working yards in the Manitowoc vicinity had also made a good name for themselves over many years of shipbuilding.

Below: **Infrastructure and tunnel works at Shakespeare Cliffs, Dover, featuring two Manitowoc Vicon heavy duty crawler cranes in the foreground.**
Transmanch Link Joint Venture

From 1902, when negotiations began for the purchase of the yard, plans were already being drawn up to bring in orders for ships and repairs. It was in 1905 that the first steel vessel, the passenger steamer *Maywood*, was launched.

In 1904, Elias Gunnell, Charles C. West and L. E. Geer organised a separate business, The Gunnell Tool Co, to manufacture a pneumatic rivet-heating forge they had invented. They also manufactured a whole range of tools which would be required for Manitowoc Dry Docks. They went on to produce marine engines and other heavy items of machinery

Left: **The Manitowoc M250 Series 2 is a 270-ton capacity crawler crane. Manitowoc began making cranes in 1925 after the shipbuilding company had gained an interest in sand and gravel businesses, where Moore-built steam cranes using grabs were being used. Manitowoc acquired the Speedcrane Co and from 1928 began producing what had been Moore Speedcranes and adding their own designs to the line over the years. Now by adding 'Ringer' attachments, they are able to lift well in excess of 1,000 tons.** *Manitowoc*

Below: **This massive construction project connected with the oil and gas industry called for very big jack lifts to be used to hoist these huge platforms. Around them are all manner of cranes including several Manitowocs and other crawler cranes, a Grove truck crane and a tower crane.** *Bridon Ropes*

required for the fitting out of large vessels. Eventually Gunnell Tool became absorbed into the Manitowoc Dry Dock Co in July 1909.

Good luck seemed to follow West, Gunnell and Geer when the Manitowoc Steam Boiler Works was put up for sale by its owner William Hess in 1908. He, like Burger & Burger, had built up an excellent reputation, so it was a sensible and obvious choice for the three men to make their bid which closed at $162,065. The purchase included all of the land and fixed assets. The two main parcels of land were interconnected by railways which led to the dockside, from where heavy boilers were loaded into ships using a 100-ton crane derrick in the form of a shear legs, which was a permanent fixture at the site.

At the Manitowoc Boiler Works, as they were known, many large items of heavy plant for industry were manufactured in addition to boilers. They included wood pulp digesters, dredger type dippers, dryers, furnaces, ladles, vulcanizers, filters, kilns and a whole host of heavy metal fabrications.

There followed the launch of many vessels either built or repaired by the company throughout the early years of this century, including many ships ordered for both world wars, including submarines for the last war. Manitowoc underwent many changes, though always the name 'Manitowoc' was retained within the company.

It was in 1925 that what had been fundamentally a shipbuilding and heavy engineering company first turned its attention to cranes and excavators. Some years earlier, in 1916, Gunnell, West and Geer obtained an interest in a sand and gravel business which had eight cranes working at its main yard. It occurred to West that manufacturing cranes themselves would serve two purposes; one to have a new product to market which would always be in demand and the second to keep his existing machine shop facilities fully occupied.

The cranes at the gravel plant had been built at Fort Wayne, Indiana, for Roy and Charles Moore of Chicago. However, Moores had shown an interest in finding a new manufacturer for its cranes, so in 1925, West and Moores got together. Manitowoc Shipbuilders subsequently agreed to manufacture the Moores Speedcranes for Moores under the Moores patent, while Moores would purchase the cranes and pay for them under certain terms.

Manitowoc was also able to loan money to Moores Bros by way of a credit agreement. Moores patents were assigned to Manitowoc as a surety against bad debts or defaulting payments by Moores. One of the important arrangements covered by Moores patents was the sliding clutch, which allowed the machine's main and only engine to control all the functions, such as slewing, or travelling, by engaging individual clutches controlled by the operator for whatever was required, without having to rely on pairs of clutches or additional engines. Thus the sliding pinion was an important selling feature of Moores cranes.

Until Roy Moore came to Manitowoc, all the Speedcranes built were four-wheel traction steam-driven cranes with a capacity of 15 tons, or three-quarter cubic yards capacity for those equipped with excavator buckets.

Steam cranes and shovels were still popular and much sought after by the contractors of the day, well into the 1920s and in some cases later. Petrol and diesel engines were, however, beginning to take over, largely due to their relative simplicity requiring less maintenance and being cleaner. No longer would there be the necessity for a second man to stoke up the boiler, as all operations could be controlled by the one operator.

Wheel-mounted cranes were seen to be good material handlers, but not especially good for digging ditches or for setting steel. The success of the crawler track, used on tractors, showed that this device was to become equally popular on cranes and excavators. So a crawler base was designed for the Moores Speedcrane.

Moores also saw the need to design a crane which could easily be converted to use a variety of front end equipment, such as crane, dragline, clamshell, shovel, backhoe or skimmer. These types of machine were eventually to be made by all excavator manufacturers, hence the name 'universal' excavator. Roy Moore got to work on these designs as soon as he became involved at Manitowoc. The first model of which was the Moores Speedcrane Model 100 of 15-ton capacity, gas driven and mounted on crawler tracks. The first was sold to a coal company in Wisconsin.

In 1927, Moores and the Manitowoc staff introduced the cast steel car body in place of the riveted mainframe previously used. Customers were given the choice of power systems between gas, steam, diesel or electric. The Model 125 later became known as the Model 1500, and was a 20-ton capacity crane or 1¼-cubic yard excavator, later upgraded to a 25-ton capacity crane in 1936 and remodelled as the 2000.

Attempts had been made to buy out the Moores patents by Manitowoc from 1933 to 1935 without success. Subsequently, royalties were paid until the patents ran out.

During the 1930s, work began on the Model 3000. Originally known as the 150, it was a 1½ cu yd excavator or a 26-ton capacity crane. Further upgrades to this model increased its capacity, followed by an all-new Model 3500 taking it to 2cu yd capacity or 50 tons as a crane. The first was shipped in July 1938. The Model 1600 which was to follow was delivered in August 1939. Strangely, the Model 1600 was smaller, designed to handle a 1cu yd bucket or 20 tons as a crane. World War 2 brought a glut of orders for cranes, with 116 Model 2000 and 79

Left: **Night work at Nigg Yard, Ross-shire, as Manitowocs and an American help the mighty Lampson lift the massive jacket in preparation for its eventual journey out into the North Sea.** *DGS Films*

Right: **A Manitowoc crawler crane resides beside the pipes on this pipe-laying barge in the North Sea in March 1968.** *Bridon Ropes*

Model 3000 cranes and shovels, together with spares packages, being delivered to the Army and US Navy during the war. Others were designed for special operations from barges, of which 58 were ordered for use on the US Navy's floating dry docks and 25 on barges themselves.

During World War 2, Manitowoc Shipbuilders was approached by the US Navy to build submarines, something they had not encountered thus far in their operations. In fact, Manitowoc, realising the severe problems, were not at all keen. Eventually they were persuaded and after lengthy refurbishment of the facilities, paid for partially by the Navy, they set about the huge task of building these big alien vessels. West approached a number of established crane makers in the US to find suitable lifting equipment to handle the sections as they were built. The crane would be required to lift 30-ton loads at a radius of 30ft. One manufacturer was able to offer such a crane but felt it would be so large that it would be impractical to operate in the confinements of the dry dock, where many other obstacles would restrict its movement, such as overhead cranes. West decided to approach his own company's crane makers to see if the design department could produce the crane he was looking for. The eventual design was to produce two cranes on extending crawler bases and with abnormally large roller paths on which it would slew. Appropriately, the torque converter manufacturer, the Twin Disc Clutch Co, had already approached Manitowoc to discuss the use of their device in the cranes. Many operators were only happy to use steam powered cranes when high torque and precision lifting was required, but the torque converters fitted to the large diesel engines were able to offer the same confidence to the operators. The Model 3900 was designed and proved successful for the submarine builders and subsequently for heavy lifting contractors generally.

Prior to building the cranes, Manitowoc designed and built a special heavy duty crawler transporter, to enable the big sections of the vessels to be moved quickly and inexpensively within the yards.

With the huge demand for cranes following World War 2, Manitowoc realised the potential for growth in the crane market. In 1947, the first Model 4500 went to work equipped as a dragline for coal stripping operations in Pennsylvania, where it was still working 30 years later. It was equipped with a 140ft boom and a 5cu yd bucket. In 1951, a Model 4500 equipped with a high-lift shovel, also for coal stripping operations, was delivered.

In 1961, the Model 4500 was upgraded to the Model 4600, increasing its lifting capacity from 100 to 200 tons and allowing buckets from 5 to 7½cu yd to be handled. However, this model was more likely to be sold in crane form than excavator.

In 1958, Manitowoc applied its exclusive Vicon concept to the 4500. Vicon is a variable, independently controlled system of power transmission that provides the operator with infinitely variable precision control over load hoisting and lowering, while minimising clutch heat and wear. It also allows the crane operator independence in travelling, swinging and booming operations.

In 1961, Manitowoc engineers designed and tested a control torque converter for crane applications which in 1962 was applied to the Vicon system. In 1962, Manitowoc was able to offer another extremely helpful idea — that of extendable crawler tracks, which allowed for more stability and greater control of heavy loads being carried off site, while making machine transportation easier and safer.

The Ringer patented by the company was the one step towards being able to provide supercranes, in that the load capacity when equipped with the new concept would be greatly increased. The Ringer assembly could, in fact, more than double the lifting capacity of the crane. Tower booms were also to become a feature from 1968.

FLOATING, MARINE & DOCKSIDE CRANES

It is interesting to observe that in 1985 there were around 46 companies manufacturing a variety of floating or shipboard cranes, ranging from 5,000 to 8,000 tons capacity. One manufacturer, Gusto Engineering of the Netherlands, themselves built off-shore cranes such as:

Hermod	4,000 and 3,300 tons
Balder	4,000 and 3,300 tons
DB101	2,500 tons
Benbecula	350/600 tons
Thor	2,000/3,000 tons
Champion	800/1,200 tons
Orca	800 tons
Challenger	800 tons
Titan 1, 2, 3, 4	600 tons
Antei	1,600 tons
Aspolin	1,200 tons
Kuroshio	2,500 tons
Crane	800/1,200 tons

In 1995 the number of manufacturers from around the world was around 36.

The oil and natural gas industries have always been major users of cranes of abnormal size, whether they operate on land or at sea. Shipping disasters and the handling of outsized loads unable to travel far on land will also provide work for giant cranes.

One such contract came to North Devon early in 1996, when a huge floating crane was brought in, probably from the experts in north Europe (predominantly the Netherlands), to offload huge vats destined for a chipboard factory on the edge of Exmoor.

The fishing boat *Pescado*, which had sunk with all lives on board being lost, was raised from the deep as part of a police investigation into the tragedy. Heavy floating cranes were used on both attempts to raise the vessel (the first attempt was sadly unsuccessful).

On land, huge cranes of immense capacity are used daily to undertake lifts which take months of

careful planning and utilise large numbers of expert manpower. Another such lift at north Devon's Appledore Docks was when a huge Gottwald truck-mounted crane was brought in to lift the cabin and superstructure from the fitting yard up onto the completed hull of a new Royal Navy ship. Two further support cranes were used to help erect the main crane.

Left: **Rail-mounted mobile quayside gantry cranes, typical of many in use to handle the vast quantity of containerised traffic at ports around the world. In the background is a line-up of conventional dock cranes.**
Clarke Chapman Marine

Top: **Three massive ship's cranes of the overhead travelling type being used to load cargo in 1970 at London Docks.**
Bridon Ropes

Right: **Munck Loaders by NKK on the *N. R. Crump* at London Docks in 1970.**
Bridon Ropes

Left: **Munck Loader overhead cranes stack thousands of tons of timber at London Docks in October 1973, with the help of on-shore fork-lift trucks.** *Bridon Ropes*

Below: **Two huge Stothert & Pitt container-handling cranes at the Port of Manchester in 1974.** *Bridon Ropes*

Right: **A view of the *N. R. Crump*, a huge cargo vessel, with its three onboard cranes at London Docks in 1970.** *Bridon Ropes*

Below right: **Paceo-Vickers dockside cranes at Seaforth Container Base in May 1973. In the dock is the container carrier *Atlantic Conveyor*.** *Bridon Ropes*

N.R. Crump

PORT OF
LIVERPOOL
ATLANTIC CONVEYOR

Right: A variety of materials being handled by cranes in this set of photographs, starting with an articulated colour television transmission trailer weighing 16½ tons being lifted by the Port of London Authority floating crane *Ajax* for shipment to Finland on the mv *Pallas* in December 1964. *Bridon Ropes/Neil Richardson*

Below: Loading a barge with Carlton Joint tubes weighing 2½ tons in 1964, with what is obviously a crane made long before the 1960s. It could well have been steam powered. (Stewart & Lloyds were the operators.) *Bridon Ropes*

Bottom: Lifting the gearbox of HMS *Invincible* at the Vickers Shipyard. Note the lifting beam is tested to a safe working load of 160 tons. *Bridon Ropes*

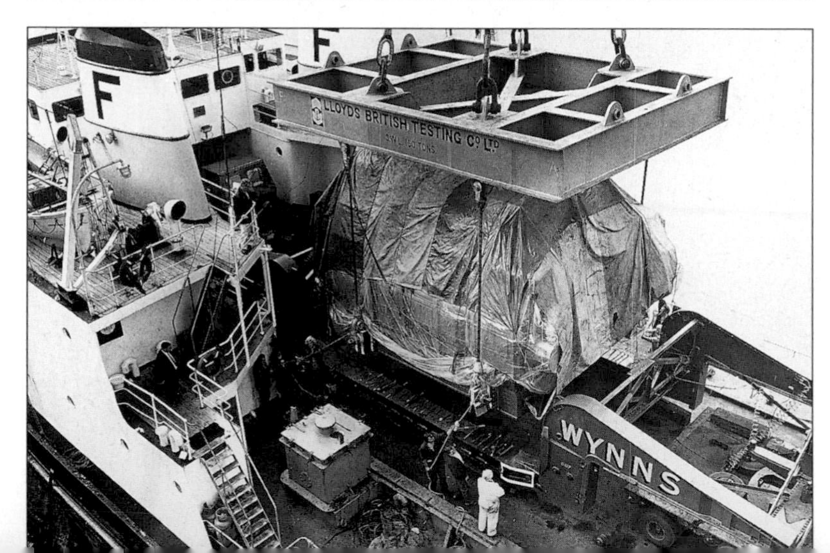

Above right: A new processing plant is hoisted into place by two large all-hydraulic truck cranes at the ICI Chemicals & Polymers complex at Middlesbrough for the contractors AMEC. *ICI Chemicals & Polymers*

Above far right: Ship's cranes, davits and derricks have been in use for hundreds of years to handle the cargo to and from ships. Here is a diagram of one such system, tracing the direction of the ropes in relation to the booms and the cargo. *Bridon Ropes/Neil Richardson*

Right: A Hallen Universal Derrick and Y-Mast on the mv *Gondul* in Millwall Docks in May 1964. *Bridon Ropes*

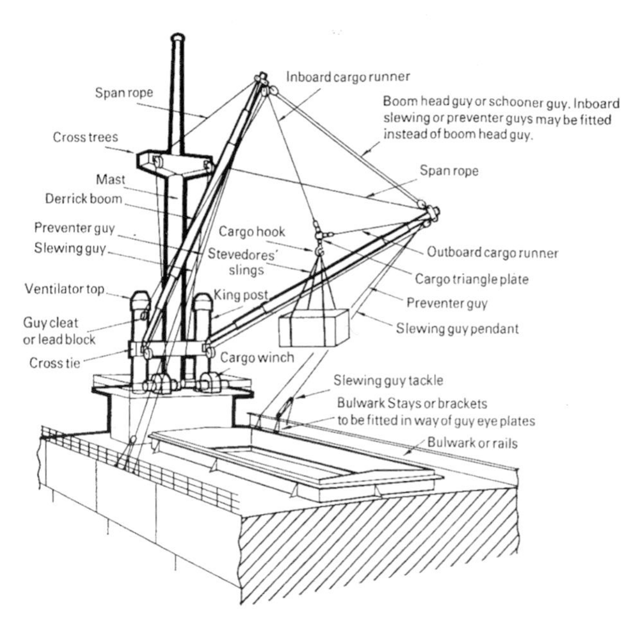
Inboard cargo runner
Span rope
Boom head guy or schooner guy. Inboard slewing or preventer guys may be fitted instead of boom head guy.
Cross trees
Span rope
Mast
Derrick boom
Preventer guy
Cargo hook
Slewing guy
Stevedores' slings
Outboard cargo runner
Cargo triangle plate
Ventilator top
King post
Preventer guy
Guy cleat or lead block
Slewing guy pendant
Cross tie
Cargo winch
Slewing guy tackle
Bulwark Stays or brackets to be fitted in way of guy eye plates
Bulwark or rails

Left: This 240-ton ammonia synthesis converter manufactured by Babcock & Wilcox is being lifted aboard the *Australia Star* at Finnieston, Glasgow, for transit to Mexico in May 1968. *Bridon Ropes*

Right: The Withy Line vessel *Newfoundland* using its ship's cranes to hoist cars aboard in June 1966. *Bridon Ropes*

Below left and below: Three photographs following the working cycle of 12-ton Velle derrick cranes equipped with grabs aboard the 20,000-ton bulk carrier MV *Brunes,* one of a series of eight vessels built by Lithgows of Glasgow for A. S. Kristian Jebsens Reden (April 1971). ***From left to right:*** Grab closing on bulk material. Hoisting the load; Entering the ship.
Bridon Ropes

Above: **Marine boatfalls, davits and rigging, together with speedcranes on board the *Malvern Prince* in May 1973.** *Bridon Ropes*

Left: **Stulcken derricks on board the Ben Line vessel *Benarty* in September 1969.** *Bridon Ropes*

Above right: **A second view of the roping on the Stulcken derricks on *Benarty*.** *Bridon Ropes*

Above far right: **A barge-mounted 'Speedcrane' undergoing testing in May 1973.** *Bridon Ropes*

Right: **Velle derrick crane in February 1977.** *Bridon Ropes*

Far right: **Electric cargo crane at the Churchill Dock terminal of Hessanatic-Neptunus NV, Antwerp.** *Bridon Ropes*

HESSE NATIE

Above left and above: Modern dockside cranes made by Butterley Engineering, at the Port of Hull. *Associated British Ports*

Left: These two cranes are somewhat unusual, one a 15-ton capacity, the other a 30-ton capacity, both mounted on the same base and slewing ring. A set of twins with a common purpose to handle cargo for the marine industry. *Clarke Chapman Marine*

Below: This crane was purpose-built for the recovery of helicopters. It is electro-hydraulic and is mounted on a dual-purpose 'solids and liquids replenishment systems' vessel. *Clarke Chapman Marine*

Above: Once again we have hybrid cranes — a 23-ton main crane and a 5-ton support with, in this case, a third crane for separate handling operations.
Clarke Chapman Marine

Above left: This 20-ton capacity electro-hydraulic crane is built to handle ship's buoys. Note its very unusual 'bifurcated' jib. *Clarke Chapman Marine*

Left: These ship's davits have but one purpose — to lower and hoist the ship's lifeboats. Nevertheless, they are cranes. *Clarke Chapman Marine*

Below: Although not easy to spot from this aerial photograph of ships refuelling at sea, there are numerous cranes aboard to help handle the big fuel lines when refuelling the aircraft carrier on one side and Royal Navy frigate on the other.
Clarke Chapman Marine

Above: **Handling timber was the task of these strange looking dockside cranes, around 1900-10.** *Associated British Ports*

Right: **These two dockside cranes are of a type familiar at dockyards, docks and ports in many countries. This is believed to be Southampton during the 1950s.** *Author/Associated British Ports*

Above: **A 71R-B electrically operated, pedestal-mounted crane working at Lea Mouth Wharf in Southampton, unloading sand from the mv *Sand Swift,* using its 4-cu yd Norton Grab. Operated by the South Coast Sand & Gravel Co, recently acquired by Hall Aggregates division of RMC, the crane and the pedestal have, in recent years, undergone extensive modifications. The crane uses a single electric motor operational system and is based on the 71R-B normally found on crawler tracks and available with a full range of front end equipment, including:- face shovel (4½cu yd capacity), dragline (normally to 4½cu yd), lifting crane (rated at 60 tons). Normal options for power were for General Motors or Cummins diesel engines, or as in this case, electric motor.** *Thanks to staff & management of Hall Aggregates for the use of this photograph*

Above: This massive crane barge operated by Brown & Root — *Atlas* — was built in the Netherlands by De Rotterdamsche Droogdok My NV and fully operational in August 1965. Note the large travelling crane overhead. *Bridon Ropes*

Below: At least two cranes operate from the BP 'Platform B' and besides the small crane to the right of the picture, we have this very large off-shore service crane operated by Brown & Root in March 1968. *British Ropes Ltd*

Right: This huge crane is operating in Sante Fe, California, in October 1973 on oilfield operations. It is called the *Choctaw* and was brought to this contract from Wilmington, Delaware. *Bridon Ropes*

Below: The *Blue Whale* is one of several very large crane barges working around the world. Some have capacities of several thousand tons. This one is no exception, being featured here around 1976. *Bridon Ropes*

Bottom: The crane barge *Thor* alongside the production platform FA (Graythorp 1) in BP's Forties Oilfield in the North Sea, after the installation of two 500-ton accommodation modules in February 1975. *Bridon Ropes*

Top right: The *Viking Piper* working for Sante Fe International Corporation on off-shore pipelaying contracts has four very large pipe handling cranes. Two of them are almost certainly Manitowocs. It is not clear about the others. Taken during the mid-1970s. *Bridon Ropes* .

Right: On board the McDermott Lay Barge from New Orleans in September 1969 we find two very large American travelling cranes mounted on rail lines to lift those pipes at any point along the barge. *Bridon Ropes*

AMERICAN
AMERICAN
McDERMOTT LAY BARGE Nº 22

Above: **The Derrick barge *Hercules* and its massive crane *Clyde,* owned by Brown & Root in the Cromarty Firth, Scotland, in July 1974. The *Hercules* is one of two specially commissioned barges which were brought in to lift into position the deck sections for the drilling and production platforms for BP's Forties Oilfield in the North Sea.** *Bridon Ropes*

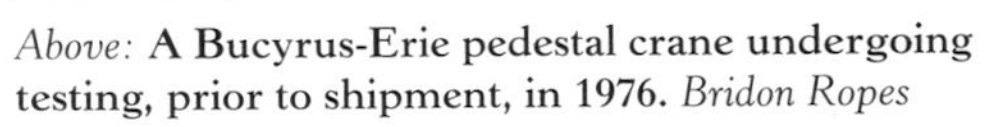

Above: **A Bucyrus-Erie pedestal crane undergoing testing, prior to shipment, in 1976.** *Bridon Ropes*

Right: **This Bucyrus-Erie crane is operational on the Ekofisk Platform in the North Sea in 1975.** *Bridon Ropes*

Below: **Also undergoing testing prior to disassembly and shipment is this Priestman Sea-Lion bound for the North Sea.** *Bridon Ropes*

Left: **The Priestman Sea-Lion 60E (electric) on the Viking Platform in the North Sea, working for Conoco in 1977.** *Priestman Brothers/Bridon Ropes*

Below: **Floating cranes are far from being a modern phenomenon. This, the *Titan 11* was working around the northeast in the early 1920s. Note the little steam crane to the right of the picture.**
Bridon Ropes/Newcastle Chronicle & Journal

Right: **This very early floating crane was known as the 'Mammoth'. It is lifting a 140-ton bridge section in 1917.** *British Ropes Ltd*

Below right: **A very modern floating crane, the 200-ton *Canute* at work in Southampton Docks during the 1990s.** *Associated British Ports/Southampton Docks*

Below far right: **An 800-ton capacity floating crane from the long established German/Austrian manufacturer Takraf, lending a hand on part of the Eurotunnel project in Dover.** *QA Photos/Eurotunnel*

TAKLIFT 1

Left: **This large floating crane was built by Blohm & Voss of Hamburg for Howard Doris of London. Known as 'Tog Mor', its capacity was 900/950 tons. The weight of the barge and crane was 4,420 tons.** *Blohm & Voss*

Below: **The floating crane *R. B. Brunel* is Britain's largest floating crane. It is being used to raise HMS *Reward*, just seen breaking the water surface.** *Bridon Ropes*

Above: **A major salvage operation in the waters off Singapore calls for very heavy lifting tackle. These take the form of the *Asian Hercules*, another floating crane at the stern end of the ship and an equally large strut jib crane.** *Bridon Ropes*

Right: **The new 35-ton capacity crane built by BM Titan at a cost of over £1 million, was designed to lift increasing amounts of heavy projects and out-of-gauge cargo.** *Associated British Ports*

Above and below: The floating crane *Ugland* is helping the tower cranes on these very big construction projects. With its outstanding reach, it is able to lift those concrete segments up onto the towers with relative ease. A whole host of dockside and floating cranes are seen in two views, including the new crane waiting to be rigged with ropes on the oil production platform being constructed at this Norwegian yard. You have to have a bit of a head for heights to operate that extremely tall dockside crane to the left of the platform in the other view. *Bridon Ropes*

Right: **Three Kone container cranes at the Virginia Port Authority in Norfolk, Virginia, USA in 1988.** *Bridon Ropes*

Below: **The huge Southampton Container Terminal uses this mighty Morris container crane to handle those big boxes. The NCK Rapier crawler crane further down the dockside ignores the goings on at this end.** *Associated British Ports/ Southampton Docks*

DERRICKS (SCOTCH TYPE)

Derrick type cranes were built predominantly in Scotland from the early 1800s (hence the description 'Scotch'). Three manufacturers are best remembered for these types of crane: Andersons, Henderson and Butters.

Butters Brothers of Glasgow (Birmingham and London), were manufacturers of a wide range of machines, including hand operated, steam and electric powered. Hundreds are still in use now in the 1990s, long after the company and its competitors have ceased making cranes.

They are used in all manner of situations, including the construction of dams, harbours, jetties and ships. When working from high steel platforms they used to perform the tasks now associated with tower cranes in the construction of tall buildings. They work in timber yards, steel foundries, quarries and have even been used to lift coal from deep opencast coal deposits.

When mounted on barges, they are able to work at sea repairing and building breakwaters and to help with sea and coastal defence work.

Today Scotch derricks can still be seen hard at work at wharfs, timber yards and quarries often more than 50 years after their manufacture and long after their original makers have ceased the manufacture of cranes.

Left: **Here is an early example of a steam powered scotch derrick. In this photograph the derrick is helping to build a bridge in 1910.**

Below: ***Weapon IV* is a barge-mounted electric derrick crane which is used daily by the sea defence and marine engineering contractor Samos of Plymouth, Devon, which uses the crane to carry out repairs to the famous breakwater in Plymouth Sound, and repairs to the harbours and docks in and around Plymouth and the Southwest. Occasionally the derrick is assisted by a crawler-mounted 22 R-B crane.**

Above: **A Butters derrick crane, which was manufactured in 1977, at the Hantergantick Quarry, St Breward, Cornwall. Its capacity is 20 tons on a jib 130ft long.** *Helen Ferrett/Natural Stone Products*

Far left: **Here, a rail-mounted electric derrick is being used to drive piles on a construction project.**
Photo courtesy of J. Laing & Co. Archivist Mr Thorpe

Left: **At the Natural Stone Products De Lank Quarry is an Anderson-Grice derrick crane manufactured in 1963. With a SWL (safe working load) of 25 tons, it operated a jib 125ft long.**
Helen Ferrett/Natural Stone Products

Below: **One of many thousands of scotch derricks which were made by the three most famous companies — Butters Bros, Hendersons and Anderson-Grice. The crane featured is one from the latter.**
Low & Duff Developments Ltd

OVERHEAD & TRAVELLING CRANES

Above: A cradle sling manufactured by British Ropes of Doncaster is being used to lift hot square steel bars from the transfer bogey at Peach & Tozer's Rotherham Steel Works in March 1964, then still a branch of the United Steel Co. *Bridon Ropes*

Top right: This overhead travelling crane is using Edge's chain slings at the Fox's Works, Stocksbridge, near Sheffield in 1964. *Bridon Ropes*

Right: Heavy duty overhead cranes are always kept busy at the South Ropery in Doncaster, where British Ropes make hundreds of miles of wire ropes for a wide range of applications, including cranes and excavators, bridges, ships, winches and transporting timber from forests. October 1973. *Bridon Ropes*

Far right: Superslings in use at the Stanningly, Leeds, works of George Cohen Ltd, the famous 600 Group of Companies, who once acquired Jones Cranes of Letchworth & Compton, c1966. *Bridon Ropes*

LIMITED
ROPES
BRITISH

SMITH - KEIGHLEY LOAD 10 TONS
BARRETT
BRADFORD
BARRETT
BRADFORD
BARRETT
BRADFORD

Left: One of the oldest manufacturers of overhead travelling cranes is Smith of Keighley, Yorkshire. This is one of their 10-ton capacity cranes in the steel girder stockyard of Henry Barrett & Sons, Bradford, in April 1964. *Bridon Ropes*

Below left: A clear view of the operator at the controls of an overhead travelling crane, using a lifting beam to stabilise this 1½-ton load of steel piping in April 1964. *Bridon Ropes*

Above: Two rail-mounted mobile gantry cranes are using magnetic and claw pick-up grabs to handle scrap at this metal merchant's yard. The cranes were built by Clyde Crane & Booth. The first one was a 7½-ton capacity crane, while the crane in the background was rated at 10 tons. *Clarke Chapman Marine*

Above right: This huge overhead travelling crane, rated at 60-ton capacity, is being used to lift tunnelling segments for the Channel Tunnel project. *Bridon Ropes/TML Eurotunnel*

Right: A 55-ton capacity Monotower crane at the Hebburn Shipyard of Hawthorn Leslie (Shipbuilders). This very large crane was built by Clyde Crane & Booth, probably some years before this photograph was taken in July 1966. Note the very big tower crane to the left of the picture. *Bridon Ropes*

Above: **A 10-ton electric overhead travelling crane, made by the Standard Steel Co (1929) and erected at Metal Propellers Ltd in Croydon.** *Low & Duff Developments Ltd*

Below: **An Anderson EOT crane with a maximum lifting capacity of 40 tons on its main hoist and a further five tons on an auxiliary hoist. This crane was made in 1956.** *Low & Duff Developments Ltd*

RAILWAY CRANES – STEAM AND OTHERS

Above: A Ransomes & Rapier 40-ton capacity steam Stokes type railway breakdown crane, *en route* to the Gold Coast Railway in 1955. *Colin Chambers*

Below: A Ransomes & Rapier 40-ton Stokes type steam railway crane at the company's Waterside Works in Ipswich, where it was made in 1955, before being transported to Nigerian Railways (3ft 6in gauge). *Colin Chambers*

Far left: Another Ransomes & Rapier steam railway breakdown crane ordered by Nigerian Railways in the 1950s. *Colin Chambers*

Left: This Ransomes & Rapier Stokes type railway breakdown crane is a 75-ton capacity machine. In this photograph we can see its unique tilting mechanism. This crane was awaiting delivery from the Ipswich Works in 1957. *Colin Chambers*

Below left: A unique photograph of a Ransomes & Rapier steam railway crane, minus its boom and rigging. It does however show the gearing. *Colin Chambers*

Right: Ransomes & Rapier DS1580, at work helping position a footbridge in the Exeter area of Devon in the 1950s. *Colin Chambers*

Below: Again in the Exeter area, DS1580 is helping a sister crane to deal with this steam railway engine in the 1950s. *Colin Chambers*

Above: DS414 (Taylor & Hubbard No 1595 of 1948) backwards in the dirt; notice how the jib is bent. Fortunately a spare jib was located at Swindon and No 414 was returned to traffic. This was surprising as a new fleet of diesel-hydraulics was shortly introduced, although teething problems caused late delivery. DS414 remained in service until 31 December 1980. Eastleigh PAD July 1978. *A. W. Chapple/Les White collection*

Above right: DS1997 (Taylor & Hubbard No 1538 of 1945), the other long time resident of Eastleigh PAD. This was to remain in service until 1980, but was withdrawn on expiry of its current certificates; however it was not cut up until September 1982. *Les White*

Below: DS58, newly restored, working at Ropley, Mid Hants Railway, c1989. Originally allocated to Woking PAD, in later years kept as spare and became semi-derelict. Sent to Eastleigh for storage about 1979, although not actually withdrawn until 1980. Purchased along with DS414 in 1981 and sent to Mid Hants Railway. *Les White*

Right: Crane lifts crane. DS58 lifting the Isles steam crane of c1896 (Higgs & Hill Ltd) for transport away from Mid Hants Railway. Now at a preservation scheme at Northampton. Seen at Alresford c1989. *K. Bynam/Les White*

Below: A better view of the Isles crane. Although it had worked at the MHR in the early days, the boiler was only on loan, hence the lack of a boiler when it was removed from railway. *K. Bynam/Les White*

Bottom: Isle of Man No 3, an early Taylor & Hubbard restored in 1994. *Les White*

Above right: **A Smith steam crane from Thos Smith & Sons, Rodley, near Leeds, is now on show at a museum near Carmarthen in Wales.**

Centre right: **Jeremiah Balmforth and Jeremiah Booth were once partners in the company of Thos Smith of Rodley. They then left to produce their own steam cranes. This shows one of the few remaining examples of a Balmforth steam crane, not only still in existence but still working at an open day at the Gloucester Waterways Museum.**
Tony Swaine/Gloucester Waterways Museum

Below: **A Cowans Sheldon railway breakdown and civil engineering crane.**
Clarke Chapman Marine/Prof F. W. Hampson

Gottwald began crane building with their first railway crane in 1906 at the MUK facilities in Düsseldorf-Reisholz, Germany. Gottwald became a majority shareholder in 1918.

The earliest cranes were built predominantly as yard cranes and were steam powered, mechanically operated and were produced until World War 2. Gottwald started the development and production of railway breakdown cranes in the 1930s.

In the 1950s and early 1960s, steam-operated railway cranes were being supplied by Gottwald to Rhodesia (now Zimbabwe), India and to the German Federal Railway (44 units of various capacities from 5 to 75 tons were supplied to India, with many of them still in operation well into the 1990s).

From 1960 to 1965 some diesel-electric railway cranes had been developed. The first railway crane to feature diesel-hydrostatic drive was built in 1960.

In the early 1960s the diesel-mechanical (also with torque converter) and diesel-electric versions dominated the market until the hydrostatic drive system was improved to the extent that it was applied for all robust moving equipment. As to railway cranes, the new system became predominant only in Germany. In 1974, however, the Danish State Railways placed an order for two track maintenance cranes of 25-ton capacity at 9m radius, to work under overhead electric power catenary in horizontal jib positions. These cranes featured hydrostatic drive.

In close co-operation with the German Federal Railway, Gottwald developed and manufactured a fully hydrostatic track maintenance crane of 20 tons capacity at 8m radius in 1978, to be operated under the catenary. With the load at the hook, the wheel load did not exceed 17 tons when travelling.

The axle load compensation in train formation (hauling speed 100km/h) was achieved by shifting forward the complete superstructure including the jib and counterweight. A high-tech railway crane with telescopic boom for track maintenance which required no rigging time was available for the first time.

In almost 20 years of operation not even a minor problem has occurred.

Those cranes and similar versions are now working in Egypt, Algeria, Denmark, the Netherlands and Germany.

In 1968, Gottwald introduced the 45-ton track maintenance crane. This particular crane was designed and developed to lift switch panels of 24-metre length and 17 tons in weight. Technical performance was outstanding. Orders were immediately placed by customers in Germany, Denmark and Switzerland. The 45-ton crane version incorporated several unique ideas like 2m tail radius and a track levelling device to compensate super-elevated curves up to 160mm. These features had been developed by Gottwald in 1984 for two 100-ton railway cranes for Belgian National Railways.

In October 1988 Gottwald was taken over by Mannesman-Demag, which considerably improved Gottwald's financial strength and contributed to the extension of Gottwald's position as a market leader in this field.

In 1989, with wooden sleepers being replaced more and more by concrete, Gottwald was asked to develop a crane which could handle switches with 24m lengths and 28 tons in weight for turnouts with a 1,200m radius.

In 1990 Gottwald moved to new premises with extended facilities and access to the workshops of Demag's other companies, for increased productivity in times of peak workload.

In 1991, Gottwald delivered the first of two units of the new 80/100-ton track maintenance crane. This machine was immediately accepted by the market as it offered such outstanding features as telescopic counterweight, track levelling device,120km/h hauling speed and others.

Further orders were received in the 1992-5 period which included 10 100-ton telescopic cranes for the Egyptian National Railways.

Gottwald was the first foreign company to manufacture and deliver a track maintenance crane to the Japanese Railway Co. This particular crane also included unique features, such as a slewing counter-weight which could be kept parallel to the track when slewing the boom, thus not hindering traffic on the adjacent line. It has a self-propelled speed of 40km/h and operates freely on rail on narrow gauge track.

In 1995, the constantly increasing demands for speed, safety and environmental-friendliness have led to the development of a completely new generation of track maintenance cranes, which are subject to a number of patent applications.

The Model GS150.24T can handle parts at a distance of up to 20m. Of course, the crane can travel with these loads in order to transport them from an assembly place to the job site.

The GS150.24T has no tail radius. It can work with full lifting capacity both in track direction and to the side without any counterweight fouling the adjacent track. Thus maximum safety is achieved for any passing train.

Lifting capacities to the side can be increased when using an outrigger. Operation of the outriggers is controlled from inside the crane cabin. The large-dimensioned outrigger plates fit into the clearance profile during transportation. Together with the very long vertical way of the propping cylinders, they avoid packing with ties in most cases and thus save time.

Unlike any other crane, the GS150.24TR only requires one outrigger to be set at a time, ie the one towards the slewing direction of the boom.

The working direction of the crane is changed without slewing the superstructure, as the boom can

Above: Gottwald railway cranes, Model GS 100.08, as ordered by the Egyptian National Railways for work in Cairo and elsewhere in Egypt. An order was received by Gottwald of Germany between 1992 and 1995 for 10 such cranes, each with lifting capacities of 100 tons. *Gottwald*

Left: A Gottwald GS 60.13 railway crane at work in Bangkok, Thailand, for the State Railway of Thailand. *Gottwald*

be telescoped in or out at either end of the crane. By the end of 1997, two such units will be in operation.

In 1995, a reorganization within Mannesman-Demag group was decided. The activities of the various companies involved in material handling which comprises the production of different types of cranes were concentrated in just one company: Mannesman-Demag Fordertechnik. Thus since 1 January 1996, the new name is: Mannesman-Demag Fordertechnik AG Gottwald.

The major project in 1996 was the finalisation of the negotiations with Indian Railways regarding the supply of 12 units, each with a 140-ton capacity breakdown crane.

Above: **A GS 150.14TR railway crane from Gottwald, in Weiss, Crailsheim, Germany.** *Gottwald*

Below: **A Model GS 80.08TR track maintenance crane, being demonstrated to potential buyers from Japan and other nations in the Far East and elsewhere.** *Gottwald*

CRAWLER & TRUCK-MOUNTED CRANES

Left: **The Coles EMA, manufactured from the early 1930s, was one of the biggest success stories in the history of Coles Cranes, ordered in their thousands by the Ministries for use by the British forces and by civilian crane hire companies alike. This particular Coles EMA was named Excalibur, and was part of the growing fleet of cranes of various makes which made up the Ainscough Crane Hire Co's machines, which now have capacities from around 5 tons to approaching 1,000 tons.**
Martin Ainscough, Ainscough Crane Hire

Below left: **This crawler crane, an NCK-Andes C41, has ended up on its counterweight because the boom was raised too high and the crane mats were too far apart.** *Tony Swaine*

Above: **Cranes galore in this photograph of the early stages of the Canary Wharf development in London's Dockland in 1986/87.**
Tony Swaine

Above right: **A 22 R-B afloat; a tracked version working on a pontoon on the Grand Union Canal at Soalbury between Milton Keynes and Leighton Buzzard.** *Tony Swaine*

Right: **An NCK-Andes being loaded aboard a barge at Brighton Marina in 1978.** *Tony Swaine*

Above: An NCK-Pennine crawler crane spreading sand for a football pitch, using a clamshell (grab). *Tony Swaine*

Below and below right: Casagrande started producing hydraulic crawler cranes in the late 1970s. Previously they had been manufacturing hydraulic powered, front-end piling rigs but had been experiencing difficulty in finding suitably powered hydraulic cranes from the established manufacturers, which had both the capacity and the robustness enough to withstand the rigours of pile driving. The first crane manufactured was a C50 (a 50-ton capacity machine). Further units from 20-ton to 90-ton were subsequently introduced. All models were in a 'C' series (C20, C30, C40, C50, C60, C70 & C90). In the photographs, the Casagrande C60 and C90 are owned and operated by Bachy UK Ltd and are constructing a diaphragm wall in Hayes, Middlesex. *Casagrande UK*

Right: A C60 operated by the Pigott Bachy Group is lowering a personnel inspection tube into a pile at Hammersmith, London. An NCK-Rapier crane is being used to auger out the holes for the piles. *Thanks to Michael Finch, MD of Casagrande UK, for the use of these photographs*

Above: **From Japan, the Sumitomo LS 100RH crawler crane working on a construction site in southern England, with a somewhat older British-built crane working alongside.** *Tony Swaine*

Right: This Smith lattice-boom truck-crane, mounted on a Foden chassis, was on regular duty on repair and maintenance of china clay processing plants for English China Clays, St Austell, Cornwall, and at their plants in Devon. ECC also used many Smith-made excavators in the extraction of waste. Most however were the tracked version Smith 21, operating either as face-shovels or draglines. *ECC International*

Below: One of the top-of-the-range, new generation NCK-Rapier crawler cranes, which these days go to 300-ton capacity, in line with some of the large cranes from the USA, Japan and Germany.
Author/Baldwins/NCK

Below right: This machine was built in Spain by the Koehring concessionaires Kynos, where Koehring models were produced between 1952 and 1974. On this occasion the crane is equipped with a diaphragm grab for accurate vertical excavation of foundations. In the background is a somewhat older mobile crane, similar to those made by Taylor of 'Jumbo' fame; however that is not to say it is from that company on this occasion. *Author's collection*

TOWER CRANES

Tower cranes have helped build some of the world's tallest structures for over 100 years. If the building has held the record for being the tallest, then it follows that the tower cranes which helped construct it are also record breakers. The World Trade Center in New York (1967/8) was built to a height of 1,350ft (412m), having 105 floors. The tower cranes which helped build this massive building were manufactured in Australia.

The twin tower buildings in Kuala Lumpur City Centre in Malaysia are now even higher than the World Trade Center, standing at 450m high.

The majority of these cranes are powered by electricity, only in certain cases do they incorporate on-board diesel engine drive.

The load range is as vast as the available height of tower cranes, from 1-ton capacity to certain cases in excess of 400 tons. Many of the larger models are

Left: **This familiar-looking tower crane is thought to be a BPR-Cadillon, made in France, lifting a precast concrete beam on this building site in December 1989.** *Bridon Ropes*

usually reserved for work in shipbuilding yards or in power stations.

They are now manufactured in 16 or more countries, with around 77 manufacturers often with as many as 20 or more models available. Certain companies are able to offer as many as 40-plus cranes. They include those with the horizontal jib, along which travels a trolley carrying the lifting hook and tackle. Or they are of the luffing type, whereby the load is raised from a sheave at the tip of the jib in the same way as with other types of crane.

The mountings can also be varied including: railway-mounted, crawler track, *in situ*, lorry-mounted or wheel-mounted. They are also available as an internal climbing type, where the usual thing is to site the crane in the lift shaft of a tall building until the construction has reached its peak. Or the external type, where the crane grows up with the building and is mounted on the outside and secured to the building at various stages to the top.

Here are a few examples of the many and various tower cranes working around the world.

Far left: A Coles Centurion truck-mounted tower crane on test at Sunderland with the first tower rig for this crane, in 1965. The crane's height was 150ft with another 150ft reach. *Author's collection*

Left: This tower crane is one of several models made by Henry Cooch of Kent. These cranes are self-erected within about 5min. It is mounted on a heavy duty 4-wheeled trailer which uses an Ackerman type steering system. The crane jibs can be luffed to an angle of 15° or 30° to enable the crane to gain extra height and clearance past large or tall buildings nearby.

Bottom left: The tower crane made by Henry Cooch, in travelling mode. *Henry Cooch Ltd*

Right: A Pingon tower crane Model 30-36 at work in the Netherlands in 1972. *Vanson Cranes Ltd*

Below: Two large Liebherr tower cranes are in place for the construction of the Kylesku Bridge in Scotland in 1982. *DGS Films Ltd*

Above: The unusual thing about these tower cranes is that they appear to be operated from down on the ground by remote control panels. *Bridon Ropes*

Left: Liebherr tower cranes. Note that the cranes on the right of the basement are operating on railway lines, while those on the left are mounted on self-propelling crawler tracks. *Bridon Ropes*

Above left: Three heavy duty tower cranes on a dam construction project. These are free standing and rail-mounted. Note how the two tall cranes to the right even have a separate crane standing at the rear of the crane, near the counterweight, in order to lift the electric motors and machinery on and off during repair or when erecting or dismantling the crane. *Author's collection*

Left: These cranes, towering into the sky of their home country in the 1980s, were made by the Simma Co of Italy. *Bridon Ropes*

Right: Liebherr tower cranes on another construction project in a European city. *Bridon Ropes*

Above left: **A luffing jib tower crane on display at a construction equipment show.** *Vanson Cranes*

Above: **Two giant external climbing tower cranes are seen on this financial institution office block.** *Bridon Ropes*

Right: **On this occasion internal climbing tower cranes invade the sky from within the building's lift shafts. The operators belong to the Mile High Club.** *Bridon Ropes*

Left: **A Munster A30 crawler-mounted tower crane, on the move between jobs.** *McCarthy & Stone*

Above: A city scene in Sweden, where a Tornborg S40/30 crane is the ideal machine for working between tall buildings, with its ability to reduce the radius of the jib while still working to capacity. *Reidar Lindstrom/Tornborg*

Below: Two Tornborg S40 folding jib cranes work alongside two conventional tower cranes of the saddle jib type, working in Sweden as they are in many countries around the world. *Reidar Lindstrom/Tornborg*

Above: This Tornborg folding jib crane is the Model S40/30 at work on a construction project in Sweden. It has a capacity of 40 tons; the '30' indicates the length of the jib (in metres). The original S40 had a jib length of 25m when designed in the 1960s. *Reidar Lindstrom/Tornborg*

THE SOBEMAI – A MULTI-PURPOSE CRANE DESIGN

One of the most unusual cranes to enter the industry in recent years is the extremely versatile Equilibrium crane from Sobemai, which was first developed from its prototype in 1979, made entirely from spare parts from a wide range of other machines.

Its capacity is 3 tons at a 15m radius. It is available with a full range of undercarriages, including rail-tracked and wheeled or as a free-standing pedestal. Alternatively, it is also suitable for mounting on board ships for use as a dredger or for cable laying.

What makes it different from most conventional cranes is its ability to balance its load (whether being used as a crane or when using a variety of grabs or electro-magnets), by the use of its 'equilibrated' moving counterweight, linked to the boom of the crane by means of a parallelogram torsion bar linkage. More recent models have extended the capacity of these cranes to 35 tons.

From these photographs of the Sobemai, which is made at factories in Belgium, it is possible to evaluate its versatility.

Left: **A rail-mounted, high mounted railway Sobemai crane of 6 tons capacity at 26.3m radius, the Type E6/26. Note as with all of these cranes, that wire ropes are no longer used to hoist the load. The entire operation is undertaken with full hydraulic control of the lifting end of the boom.**

C.R.I. FRANOIS

Sobemai
ragg eisen

Above left: The perfect crane for a large, busy scrap yard with its outstanding reach and precision hoisting and lowering of its load. This crane appears to be rail-mounted.

Left: Once again in a scrap yard, this Sobemai dwarfs all that surrounds it, including the railway trucks beneath its extended lifting arm, equipped with hydraulic tine grab. *Dell Crane Services*

Above: This crawler-mounted Sobemai equilibrated crane has a capacity of 15 or 25 tons, has a reach of 38.2m and is mounted on three crawlers (2+1). *Syd Wilson/Wilson Industrial Services*

Above right: A track-mounted Sobemai using a grab to handle bulk materials, possibly cement at this port. Note its height, shown by the men standing beneath it. *Wilson Industrial Services*

Centre right: Employed by C. R. Massie in Liverpool, a special quay crane operated to suit the dock. Its capacity is 10 tons over a 25m radius. Its slewing ring is 11m above the quay. It is using a 6-tine, 3.5cu m scrap grab which has an output of 350/500 tons/hr. The grab tine tips are able to operate 8m below the level of the quay. The crane is electrically powered with 145kW in total. *Wilson Industrial Services*

Right: This crawler-mounted grab crane is operating in yet another scrap yard. *Dell Plant & Crane Services*

The wheel loader running beneath the Sobemai grab crane has no trouble finding its way past those huge crawler tracks. *Wilson Industrial Services*

COOCH SMALL MOBILE CRANES

Only Jones Cranes and R. H. Neal were well known during the 1950s for producing the little 15cwt capacity self-contained mobile cranes, until Henry Cooch Ltd started producing this type of crane during the early 1960s.

Many of the Cooch models are rated at 1-ton capacity and are frequently used in and around London to repair the sewer systems and underground services, by lowering equipment and buckets through holes in the road. What is different about these little cranes is the use of hydraulic cylinders to raise and to lower the jib. Hydraulics are also used to power the hoist drum and to drive the machine, using two hydraulic motors to drive the rear wheels. Steering is also accomplished by hydraulic cylinders.

Right: **The JC 20, 1-ton capacity mobile crane.**
Henry Cooch Ltd

Below: **This JC 20 is at home in a builder's yard.**
Henry Cooch Ltd

Above: **The JC 60 hydraulic crane, with telescopic boom and 3-ton lift.** *Henry Cooch Ltd*

Left: **Another look at the JC 20 mobile hydraulic crane.** *Henry Cooch Ltd*

Below: **The JC 60 is a fully mobile telescopic crane, using full hydraulics and a telescopic boom. It is rated at 3 tons capacity.** *Henry Cooch Ltd*